About Island Press

Island Press, a nonprofit organization, publishes, markets, and distributes the most advanced thinking on the conservation of our natural resources— books about soil, land, water, forests, wildlife, and hazardous and toxic wastes. These books are practical tools used by public officials, business and industry leaders, natural resource managers, and concerned citizens working to solve both local and global resource problems.

Founded in 1978, Island Press reorganized in 1984 to meet the increasing demand for substantive books on all resource-related issues. Island Press publishes and distributes under its own imprint and offers these services to other nonprofit organizations.

Support for Island Press is provided by the Geraldine R. Dodge Foundation, the Energy Foundation, the Charles Engelhard Foundation, the Ford Foundation, Glen Eagles Foundation, the George Gund Foundation, William and Flora Hewlett Foundation, the James Irvine Foundation, the John D. and Catherine T. MacArthur Foundation, the Andrew W. Mellon Foundation, the Joyce Mertz-Gilmore Foundation, the New-Land Foundation, the Pew Charitable Trusts, the Rockefeller Brothers Fund, the Tides Foundation, and individual donors.

About The Wilderness Society

The Wilderness Society is the only national conservation organization that is devoted primarily to the protection and management of public lands. Founded in 1935 by two foresters, Bob Marshall and Aldo Leopold, the society uses a combination of advocacy, analysis, and public education in its campaigns to improve the management of America's national parks, forests, wildlife refuges, and Bureau of Land Management lands. The Wilderness Society is headquartered in Washington, D.C., and currently has over three hundred thousand members nationwide.

DEFINING SUSTAINABLE FORESTRY

DEFINING
SUSTAINABLE
FORESTRY

Edited by
Gregory H. Aplet, Nels Johnson,
Jeffrey T. Olson, and V. Alaric Sample

THE WILDERNESS SOCIETY

Foreword by EDWARD O. WILSON

ISLAND PRESS

Washington, D.C. ❏ *Covelo, California*

ISLAND PRESS is a trademark of the Center for Resource Economics.

Library of Congress Cataloging-in-Publication Data

Defining sustainable forestry / edited by Gregory H. Aplet . . . [et al.] ; The Wilderness Society ; foreword by E. O. Wilson.

p.   cm.
    Papers originally presented at a conference held Jan. 1992.
    Includes bibliographical references and index.
    ISBN 1-55963-233-X (cloth)—ISBN 1-55963-234-8 (paper)
        1. Sustainable forestry—Congress.   I. Aplet, Gregory H.   II. The Wilderness
    Society.
        SD387.S87D44   1993
        333.75'16—dc20   93-8389
                                                                                    CIP

Printed on recycled, acid-free paper

Manufactured in the United States of America

10   9   8   7   6   5   4   3   2

Contents

FOREWORD

Forest Ecosystems: More Complex Than We Know

The green plants are the superstructure and energy source of the forest system, and the remaining organisms, from bacteria to fungi, arthropods, and vertebrates, are the equally vital infrastructure and conduits of nutrients and energy. Our understanding of the infrastructure is strikingly poor, so much so that a large percentage of the species in temperate forests—perhaps a majority in tropical forests—still lack even a scientific name. This diversity of life forms undergirds the stability of forest ecosystems to a degree that has thus far been only crudely estimated. In general, the maintenance of biodiversity as a factor in forest productivity and value has only begun to be studied systematically.

Consider that the number of species on earth is unknown to the nearest order of magnitude, whether closer to ten million or a hundred million. Biologists have given latinized names to approximately 1.4 million species, but it is estimated that there are at least ten times that many kinds of insects and other arthropods in the tropical rain forests alone. The diversity of visible organisms may be dwarfed in turn by the variety of bacteria: recent biochemical tests suggest the presence of several thousand species of organisms in a single pinch of north temperate forest soil, almost all of which remain unclassified by microbiologists. A large majority of flowering plants are known, but there are still places with floras waiting to be explored. Vast reservoirs of diversity also exist at the other levels of biological organization, from the letters of genetic code within local populations to ecosystems comprising large numbers of species. Almost none has been thoroughly studied. The life of this planet, in short, is largely unexplored.

It follows that our weak knowledge of diversity redounds into ignorance concerning the relation between species assemblages and ecosystem functioning. We should keep in mind that every species is part of an ecosystem, an expert of its kind, tested relentlessly by natural selection as it spreads itself into the food web. To remove it is to entrain changes in other species (raising the populations of some, reducing or even extinguishing others) and as a result to risk a downward spiral of the larger assemblage.

But downward by how much? The relation between biodiversity and stability is a gray area of science. From a few key studies of forests we know that diversity enlarges the capacity of the ecosystem to hold on to nutrients. With multiple plant species, the leaf area is more evenly and dependably distributed. The greater the number of plant species, the broader the array of specialized leaves and roots and the more nutrients the vegetation as a whole can seize from every nook and cranny at every hour through all seasons. The extreme reach of biodiversity anywhere may be that attained by the orchids and other epiphytes of tropical forests, which harvest soil particles directly from mist and airborne dust that would otherwise blow away. In short, an ecosystem kept productive by multiple species is an ecosystem less likely to fail.

If species composing a particular ecosystem begin to go extinct, at what point will the whole machine sputter and destabilize? We cannot be sure because the natural history of most kinds of organisms is not well enough known and the necessary experiments on ecosystem failure have not been done.

Think nonetheless of how such an experiment *might* unfold. If we were to dismantle an ecosystem gradually, removing one species after another, the exact consequences at each step would be impossible to predict (given the current lack of information), but one general result seems certain: at some point the ecosystem would suffer a catastrophic collapse. Most communities of organisms are held together by redundancies in the system. In many cases two or more ecologically similar species live in the same area, and any one would fill the niches of others extinguished, more or less. But inevitably the resiliency would be sapped, the efficiency of the food webs would drop, and nutrient flow would decline. Eventually one of the elements deleted would prove to be a keystone species. Its extinction would bring down other species with it, possibly so extensively as to alter the physical structure of the habitat, in the same way that the extinction of sea otters collapses the kelp forests. Because ecology is still a primitive science, no one is sure of the identity of most keystone species. We are used to thinking of the organisms in this vital category as being large—sea otters, elephants, Douglas firs, coral heads—but they could as easily include any of the tiny invertebrates, algae, and microorganisms that teem in the substratum and that also possess most of its protoplasm and move the mass of nutrients, the little things that run the earth.

What demonstrates the amount of biodiversity in the first place? A few suggestive principles have emerged on a global scale. The largest number of species occur in those parts of the world that possess the most ESA: the highest amount of solar Energy, the greatest climate Stability, and the largest geographic Area. The habitats maximizing all three of these qualities are the tropical rain forests, believed to contain more than half of all species on earth. Second in rank are coral reefs, appropriately called the rain forests of the sea. Also rich in biodiversity are the floors of the deep sea basins. These remote habitats are poor in energy but possess the other two diversity-generating qualities: they have remained relatively unchanged across vast reaches for tens of millions of years.

In biologically rich and impoverished habitats alike, small organisms are more diverse than are large organisms. There are more kinds of mice than elephants, more sparrows than crows, and apparently vastly more bacteria than trees. Small size allows organisms to increase their total living space by fractal expansion and then to divide that space into smaller niches, accommodating a correspondingly larger number of species during the course of evolution.

In conclusion, it seems clear that forestry is ultimately a branch of ecology and will advance as a discipline only as swiftly as ecology advances. Progress in both domains will depend in good part on the study of biodiversity; the variety of life in the infrastructure of forests is immense, fragile, and still very poorly understood. Finally, the two disciplines are bonded in another way: biodiversity is vital to healthy forests, while proper forest management is vital to the maintenance of biodiversity.

Edward O. Wilson

DEFINING SUSTAINABLE FORESTRY

INTRODUCTION

Defining Sustainable Forestry

V. Alaric Sample

Nels Johnson

Gregory H. Aplet

Jeffrey T. Olson

We are now in a period of unprecedented public concern over the alteration and destruction of forests and the accelerating rate of loss of plant and animal species native to forest ecosystems, in the United States as well as abroad. Many ecologists agree that neither our current system of forest reserves (such as national parks, wildernesses, and research natural areas) *nor any conceivable such system* will be sufficient to provide adequate protection of biodiversity in the wide range of forest habitats (Franklin 1989). We are urged to look beyond the borders of public forest lands, which usually do not correspond with ecological boundaries, and consider all lands within the ecosystem as important to its overall functioning and sustainability. But we are also urged to think beyond simple preservation and discover ways in which the protection of biodiversity and other environmental values can be thoroughly incorporated into the management of lands for a variety of uses and values, right up to the suburban boundary.

The shift toward ecologically based forest management regimes reflects the expanding mix of values that the American public attaches to its forest resources. Today, more emphasis has been placed on biodiversity and the hydrological, recreational, and climatic values of forests than at any time in the past. For most of this century, these values were of secondary importance to timber and other commodities. The legacy of managing forests as if they were important only for their timber has become increasingly evident around the country. A steadily growing list of threatened and endangered forest-dwelling species, endangered salmon runs and other declining fisheries, and degraded water quality is a troublesome indicator of the status of forest ecosystems and their capacity to meet long-term human needs. The virtual disappearance of old-growth forests from the American landscape is hardly a model we wish other countries to emulate, especially in the biologically rich tropics.

Widespread public concern for biodiversity and the express willingness to accept higher consumer prices to protect habitat and other environmental values suggest that this issue is not one that will fade away soon (The Roper Organization, Inc. 1992). The practice of forestry in the United States is currently undergoing the most profound and rapid change since its establishment a century ago. The evolution from sustained-yield management of a relatively small number of commercial tree species to the protection and sustainable management of forest ecosystems is changing some of the fundamental premises of forest management. Like sustained-yield forest management, which was a radical departure from current practice when it was introduced near the turn of the century, the protection and sustainable management of forest ecosystems seems destined to become the "conventional wisdom" in forest resource management in the twenty-first century (Sample 1991).

Recently, a number of events have signaled the emerging importance of protecting biodiversity in managed landscapes on both public and private forest lands. In June 1992, both the Forest Service and the Bureau of Land Management announced new policies adopting an ecosystem management approach for more than 200 million acres of federal forest land. Both agencies are now developing detailed guidelines for policy implementation that will be suited to the particular characteristics of the major bioregions of the United States. At the same time, a task force chartered by the President's Council on Environmental Quality to explore options for conserving biodiversity on private lands documented a variety of voluntary biodiversity protection measures taken by landowners, including several large forest products corporations. Finally, Agenda 21 and the Forest Principles were both adopted by the United States at the Earth Summit in Rio de Janeiro. By signing these agreements, President Bush committed the United States—in principle—to managing its forest ecosystems on a sustainable basis for all of their diverse products, services, and values.

In January 1992, the Wilderness Society's Bolle Center for Forest Ecosystem Management, American Forests' Forest Policy Center, and the World Resources Institute cooperated to bring together an unusually diverse group of organizations and institutions and convene a national conference entitled *Defining Sustainable Forestry*. This collaboration followed the independent realization by researchers at all three organizations that the concept of sustainable forests, despite its wide appeal, is not widely understood by the public, by forest policy and management professionals, or even by its advocates.

The purpose of the conference was to begin developing a common framework upon which to base the future development of forestry; the conventional approaches that have defined forestry for the past century were to be replaced by the ecosystem approach that will guide it into the next. The workshop was attended by 136 participants representing a wide range of perspectives on challenges facing forests and forestry in the 1990s. Ecologists, foresters, economists, and sociologists all had a chance to present their views in five panel discussions. More importantly, conservationists, public and private forest managers, policy makers, and forest industry representatives all found themselves together in working groups debating the issues and exploring for common ground. Several key findings emerged.

First, we are still so lacking in our understanding of the functioning and response of forest ecosystems that scientists are unable to articulate definitively how an ecosystem approach will look different from conventional forest management practices. The National Research Council, a part of the National Academy of Sciences, recently assembled a committee of eminent natural scientists from universities and other research organizations across the country to assess the status and needs of forestry research. Their report, entitled *Forestry Research: A Mandate for Change*, concluded the following (National Research Council 1990):

> Although concern about and interest in the global role and fate of forests are currently great, the existing level of knowledge about forests is inadequate to develop sound forest management policies. Current knowledge and patterns of research will not result in sufficiently accurate predictions of the consequences of potentially harmful influences on forests, including forest management practices that lack a sound basis in

biological knowledge. This deficiency will reduce our ability to maintain or enhance forest productivity, recreation, and conservation as well as our ability to ameliorate or adapt to changes in the global environment.

The committee called for a fundamental redefinition of forest science. The central recommendation of the report was to broaden forestry research from the agricultural model of simply improving the production of commodities to one based upon gaining a better understanding of the functioning of healthy forest ecosystems, what the report called an "environmental paradigm."

Since publication of the NRC report, increased effort and resources have been dedicated to improving our understanding of the natural functioning and response of forest ecosystems, but research findings to date are simply insufficient to indicate precisely how sustainable forest ecosystem management practices should evolve from conventional sustained-yield forest management. We can be sure only of the general direction in which we should head, and we must take special care in the meantime to avoid near-term resource management decisions that may overly restrict or foreclose future management options.

Second, an ecosystem approach will require not only the consideration of a broader array of species within the management unit but also a broader concept of the management unit itself—landscape-scale areas defined along ecological boundaries and, in most cases, encompassing a host of both public and private forest lands (Franklin 1989). It is clear that we need to look beyond just the commercial tree species and consider the broader array of plants and animals that make up the forest ecosystem, but ownership boundaries are legal or political boundary lines that seldom have any ecological basis. A more enlightened approach to the management of one parcel of land may be of little consequence if there were no consideration of what was taking place simultaneously on other parcels within the same ecologically defined boundaries. During the conference it became clear that, if there was to be any coherency in the establishment and achievement of ecosystem management goals, a far higher level of cooperation, coordination, and collaboration would be needed among the various public and private landowners and managers occupying a given ecologically defined management unit.

Third, it is clear that, once we have the science of ecosystem management well in hand, we will still have to address an array of social, economic, political, and institutional factors that will profoundly affect our ability to implement an ecologically sound approach to forest management. One significant institutional and policy challenge, alluded to above, is to facilitate closer cooperation, coordination, and collaboration among adjacent public and private forest landowners in the establishment and achievement of ecosystem management goals. Another challenge is the reorganization of resource-managing organizations, both public and private, away from function-based, target-oriented hierarchies toward open organizations conducive to multidisciplinary approaches to achieving desired future resource conditions. The social, political, and economic implications of such developments could be quite positive, but this is still unclear, and the uncertainty is troubling to many people and institutions critical to the successful implementation of an ecosystem approach to forest management.

Finally, and perhaps most importantly, an ecosystem approach must be not only ecologically sound but also economically viable and socially responsible—if it is lacking in any one of these three areas the system will collapse. Ironically, this is a lesson that conservationists learned first in the developing countries, where nature preserves drawn on maps were often overrun by local communities whose needs for subsistence overwhelmed attempts to protect wildlife or native rain forests. Each of these three considerations represents a circle of possible options; where all three circles overlap with one another describes the subset of options that define sustainable forestry. Others view this as an analogue to the "fire triangle" (fuel, oxygen, heat)—all three factors are needed to establish and maintain a sustainable system. A focus on only biophysical factors, with little or no consideration of social and economic needs, is doomed from the start.

This book captures the essence of the workshop through the rearticulation of oral remarks made by the authors at the workshop. All of the authors, with the exception of Michael Toman, actively participated in this gathering. The chapters that follow are organized into three sections. Part I briefly describes the challenge of sustainable forestry and considers some of the issues inherent in deciding what should be sustained and for whom. Part II describes many of the ecological and silvicultural aspects of a sustainable forest management system. Part III addresses some of the economic, social, political, and institutional factors that are equally critical in a truly sustainable system. It was not our intent to provide the last word on defining sustainable forestry. Indeed, at this early stage in the evolution toward managing forests as diverse, naturally functioning eco-systems, we sought only to begin the dialogue, knowing that many others would follow to fill out the framework that has been set up here.

This unprecedented workshop was made possible through the generous financial support of the Surdna Foundation, the Environmental Protection Agency, and the Pew Charitable Trusts. In addition, a broad range of co-sponsors helped to generate widespread interest in the workshop and helped in a variety of ways to make the workshop a success. The co-sponsors included the U.S. Environmental Protection Agency; USDA Forest Service; Yale University School of Forestry and Environmental Studies; Purdue University, Department of Forestry and Natural Re-sources; University of California at Berkeley, Department of Forestry and Resource Management; Environmental Defense Fund; Resources for the Future; Pinchot Institute for Conservation; and National Woodland Owners Association. We are especially grateful to all of the participants at the workshop, who—each and every one—shared their best thoughts and ideas with us and everyone else present at the workshop. Finally, we owe a considerable debt to the authors, who not only stimulated the thinking of everyone at the workshop but also made the publication of this book possible.

REFERENCES

Franklin, J. F. 1989. Toward a new forestry. *American Forests* 95:37-44.
National Research Council. 1990. *Forestry research: A mandate for change.* Washington, DC: National Academy Press.
The Roper Organization, Inc. 1992. *Natural resource conservation: Where environmentalism is headed in the 1990's.* Washington, DC: Times Mirror Magazines Conservation Council.
Sample, V. A. 1991. *Land stewardship in the next era of conservation.* Milford, PA: Corey Towers Press.

PART I

Sustain What? Exploring the Objectives of Sustainable Forestry

Introduction

Nels Johnson

World Resources Institute

During the past decade, sustainability has become the central concept invoked in the pursuit of redefining human relationships with nature. To be sustainable is the goal of a growing list of human activities that place demands on the environment—agriculture, fisheries, energy production, economic development, and, certainly not least, forestry. If only we could do these things right—sustainably, that is—society could get what it needs from the environment now while leaving nature's diversity and productivity unimpaired for future needs. Virtually everyone is for sustainability in much the same way that nearly everyone is for justice. The notion of sustainability is attractive, in part, because it seems to offer a middle ground between choosing extremes: ill-informed, inefficient, short-term, and unsustainable consumption on the one hand and no consumption at all (sustainable for the environment perhaps but not for humans) on the other.

Earlier in this century, spurred in part by the destructive practices of the timber barons, American foresters developed sustained timber yield principles based on the "scientific forestry" first developed in Europe. For decades, these principles have been at the core of forestry education and silvicultural practice around the world. Nevertheless, from the island of Borneo to the Pacific Northwest, new debates on forest sustainability have been fueled by more sophisticated ecological and social research and by widespread evidence of the negative environmental effects of logging in natural forests. Although the most damaging evidence against current forest practices comes from areas where sustained timber yield principles have not been seriously applied, the new pursuit of sustainability is pushing aside earlier conceptions of sustainable forestry. It is no longer enough simply to sustain timber yields if it is ultimately the forest that one wants to sustain.

If sustained yield is no longer considered sufficient to sustain forests, what will replace it? For what and for whom is forestry to be sustainable? Since forests embody many things—products, services, and values—for both humans and nature, which things do we wish to sustain and why? It is here that the concept of sustainability, as it is applied to forestry, begins to encounter problems. Can forests sustain demands for more housing, more paper, and more recreation, while simultaneously maintaining biological diversity, supplying quality water, moderating climate, and supporting local and regional economies? Upon closer examination, sustainable forestry is about making choices, not avoiding them. Indeed, as the prominent tropical forester Duncan Poore noted, it is questionable whether one universal definition of *sustainable management* is useful (Poore 1989). Sustainability is in the eye of the beholder. The reality of sustainable forestry demands choosing what is to be sustained—and for whom—and moving in that direction sooner rather than later.

In this section, the authors consider some of the choices that will have to be made in developing new paradigms for the management of forests and the human uses of them. If there are no universal ideal definitions of sustainability, these authors suggest that there are certainly more optimal definitions than today's prevalent practices.

Reed Noss approaches forests and forestry with a landscape ecologist's eye for space (large) and time (long). Such perspectives can be discomforting to those whose views are shaped by concerns that are both shorter in time and less expansive in area. Noss begins by stating that trying to determine what sustainable forestry is answers the wrong question. The real issue is to find ways to sustain forests, not forestry; they are not the same thing. Forestry, he maintains, is not enough to sustain forests, with its preoccupation on the continuous production of wood and other forest products. Forests are worth more than the sum of those products and more even than the added sum of such "environmental services" as climate regulation, water supply, pest control, gene banks, or recreation. "Forests,"states Noss, "are valuable and must be sustained for their own sake."

Such views are likely to strike many as reflecting a strong biocentrist bias, to which Noss would readily agree. In a world of climate change, ozone depletion, human population growth, and pestilence, however, biodiversity is likely to be the ultimate key to sustainability, including that of the human species. A landscape with a great diversity of habitats, species, and genotypes is more adaptable to change than is a monoculture. In essence, it is sustainability that depends on biodiversity, not the other way around.

Like many others who have seriously contemplated the subject, Noss believes that sustainability is a multifaceted and relative concept. In the end, he thinks that it must be defined in relation to two elements: the time period and the proportion of ecosystem structure, function, and composition that is maintained. The path to sustainable forests can only be found by first understanding what he calls the "trajectories of impoverishment." For example, biotic forest impoverishment has been fueled by replacing older forests with younger forests; replacing complex, uneven-aged stands with simplified, even-aged stands; and reducing large forest patches to ever smaller forest patches.

To reverse these trajectories is to move toward (ecologists never move back) sustaining forests. We should thus strive for older forests, growing in larger and more well-connected patches of structurally complex stands where prescribed fires are a major management tool and roads are less common. These transitions will allow the maintenance and recovery of biodiversity and the restoration of forests where they have been seriously degraded.

With five and a half billion people on earth (and growing), Noss acknowledges the need for land management—we simply have no choice. The most important question to ask with respect to sustainability, he asserts, is about making choices: What do we wish to sustain and why? What goals do we wish to attain with our land management programs, and what values are behind those goals? These fundamental questions must be answered before we can define sustainable forestry.

Hal Salwasser's conceptual approach to sustainability borrows from the "fire triangle" familiar to most forest managers, in which fuel, oxygen, and heat are necessary to sustain a fire; take away any one element, and the fire will go out. Echoing remarks made by Gary Hartshorn at the outset of the workshop, Salwasser and his coauthors note that "to conserve ecosystems, regardless of the specific goals and objectives, management must be ecologically sound, economically viable, and socially responsible." A failure to devote adequate attention to any one of the three legs of the triangle will result in a system that cannot be sustained.

Salwasser et al. also note that although much of the debate over sustainable forestry has addressed the necessary change in focus from stands to landscapes, real sustainability must account for even larger scales up to and including the global environment. If we have instituted sustainable forestry in our own backyard—but have done so through exporting the ecological effects of our demands for forest products to nations with fewer environmental safeguards on timber harvesting—have we truly achieved sustainability? With the recognition that we live in a global economic market as well as a global environment, we have no choice but to consider the consequences of our actions at multiple scales, including the globe.

Finally, Salwasser et al. suggest some basic principles of sustainable forest management that define the approach being taken by the USDA Forest Service in the management of the National Forest System. The essence of the new approach is to conserve biologically diverse and productive landscapes within local, regional, national, and global contexts. Citing the chief of the Forest Service and noted conservationist Aldo Leopold, they conclude that sustainable forestry will require "renewed vigor in pursuit of the ideals of conservation, land stewardship, and multiple-use management" and call for renewed "individual responsibility and the sense of community among the people, land, and resources."

What motivates people and institutions to seek and accept change, particularly changes that may affect their livelihoods and futures? George Honadle reflects on the experience in the tropics with changing behavior away from overexploitative resource use to the more sustainable stewardship of forests.

Those involved in forest conservation in the tropics have addressed sustainability issues more directly and at an earlier date than have foresters in North America. In many tropical areas, conservation efforts that do not address local people's needs or acknowledge their right to survival are doomed to failure. At the same time, many people's livelihoods depend on the resources the forest provides beyond timber. Thus, the need to integrate human and societal needs to survive with actions to protect and conserve forest ecosystems has provided strong impetus for developing sustainable—or at least sustainable enough—forestry.

Despite the more widespread recognition of the need for sustainable forestry in the tropics, its practice continues to be extremely limited. The fundamental reason for this, Honadle believes, is that the problem has been oversimplified to one of inadequate information—"if only people knew what they were doing, they would stop!" In other words, the solution to unsustainable forest use is better conservation education. Honadle suggests that there is a wide gap between knowledge and behavior. The solution is most likely to be found by recognizing that people respond to incentives for behavior that enhance their chances for survival. Those incentives, not surprisingly, are created by a combination of economic market forces and bureaucratic policies, rules, and institutions.

Honadle's analysis implies that reforming institutional structures and behaviors is perhaps the most important avenue for changing unsustainable forest uses. At the very least, all paths to a more sustainable forestry cross important institutional ground. Honadle identifies a set of three key institutional imperatives that will be essential to the transition to sustainable forestry, whatever its definition.

This first institutional imperative is what Honadle calls "institutional mapping." This basically involves identifying all forestry-relevant institutions and developing, analyzing, and monitoring indicators of institutional performance on forest issues. Second is selective institutional strengthening with an emphasis on creating institutional checks and balances, promoting individual and community access to institutional processes, and favoring the development of training, protective, and monitoring functions of forestry-related institutions over exploitative functions. Last, more sustainable forestry will require institutional invention. As Honadle notes, "we use organizational mechanisms not because they are necessarily what we need, but because they are there."

Honadle concludes with the identification of seven areas of concern for the promotion of sustainable forestry in North America. Some of these needs—a population policy, for example—are commonplace considerations in the tropics, but are rarely thought of in the North American context.

The recurrent theme throughout this section is the realization that there is no one perfect definition of sustainable forestry. Part of the reason for our present forest sustainability crisis is that we have trusted one version—sustained yield—for too long and to the exclusion of any alternatives. The solutions to achieving more sustainable forests will be diverse. Sustainable forestry is likely to remain a frustrating concept without more careful consideration of what we want to sustain, for whose benefit, and why.

REFERENCE

Poore, D. 1989. The sustainable management of natural forests: The issues. In *No timber without trees: Sustainability in the tropical forest*, eds. D. Poore, P. Burgess, J. Palmer, S. Peitbergen, and T. Synott. London: Earthscan Publications.

1

Sustainable Forestry or Sustainable Forests?

Reed F. Noss
College of Forestry, University of Idaho

Sustainability is the motherhood and apple pie of modern conservation. We invoke the concept in discussions of economic development, agriculture, forestry, fisheries, wildlife management, and general relationships between human beings and nature. Sustainability is a goal that no one, from the most determined of environmentalists to the most aggressive of developers, can oppose (at least not in public). If an enterprise is sustainable, it is good. If it is not sustainable, then we had better replace it quickly with something that is. Anybody who opposes sustainability must be either evil or deranged.

Amidst all the clamor about sustainability, has anyone stopped to ask, Is it true? Can we really sustain all of those exploitative activities forever? Can we keep on cutting timber, grazing cows, catching fish, and building houses, eternally, if we do these things *right*? With five and a half billion people on Earth? The situation is particularly ironic when those of us in the rich countries tell people in the Third World that they must practice sustainable development when we have never demonstrated it ourselves. We got rich on unsustainable development.

But I should not be too cynical. Sustainable forestry, and perhaps even sustainable development, are possible in principle. The sustainability concept, however, as it has been applied so far, is utterly anthropocentric— and that is where it is in need of significant revision. We might very well sustain an economic activity for a long time but yet lose many things not of immediate concern to us in the process. Some of those things we and our descendants will never miss, for we never knew they were there. Ecosystems do contain some functional redundancy. To a point (although we do not know precisely where that point is), ecosystems can be simplified and brought under our control and yet still function in the sense of cycling nutrients and transforming energy into useful products. They might even be aesthetically attractive—a pastoral scene, for example—though impoverished in many ways that we do not see or understand.

My major point in this chapter is that sustaining forestry is not the same as sustaining forests. Forestry has been largely concerned with silviculture, defined as "that branch of forestry which deals with the establishment, development, care, and reproduction of stands of timber" (Toumey 1947). The aim of silviculture, according to Toumey, is the "continuous production of wood." But forests comprise much more than wood and other products for human consumption, much more even than the "public service" functions of climatic regulation, water supply, pest control, gene banks, or recreational opportunities. What we or future human generations can afford to lose is not the only consideration. Forests are valuable and must be sustained for their own sake. Until we acquire such an attitude, the sustainability concept may be a smoke screen, behind which we continue to chip away at our biotic heritage.

SUSTAINABILITY AND BIODIVERSITY

Sustainability and biodiversity compete for the title of "biggest buzzword in conservation." Most of us would agree that these concepts should be complementary rather than competing. But which concept is more fundamental? Biodiversity is sometimes seen as a means to an end, on the one hand as a genetic warehouse from which we can draw all sorts of useful products and, on the other hand, as an indicator of environmental health. A diverse environment is a healthy and stable one, or so the old paradigm goes. Although the diversity-stability relationship is much more complicated than we once thought, no one will deny that, when an ecosystem is impoverished (such as when a diverse forest is converted into a cattle pasture or cornfield), it is often put in a precarious condition maintained only by vast inputs of cultural energy, if it can be maintained at all.

An alternative view of biodiversity, and perhaps the unspoken majority view among biologists and environmentalists, is that biodiversity is an end in itself. We are interested in preserving the full richness of species, genetic material, and ecosystems on Earth because they have an inherent worth that overshadows any use we might make of them. The intrinsic value and instrumental value conceptions of biodiversity are not mutually exclusive. We can value forests for their own sake and use them too, but not every forest and not for everything.

Another critical consideration is that biodiversity is not just a numbers game. At a global scale, maintaining maximal species richness is a very important goal. At any smaller scale, however, quality is more important than quantity. Increasing the sheer number of species in a landscape, as we can do quite handily with checkerboard clear-cutting, for example, does not necessarily contribute to biodiversity conservation goals. In fact, as we artificially increase species richness at a local scale by favoring weedy, opportunistic species that thrive with human disturbance, we risk depletion of global biodiversity as sensitive and endemic species are lost and every place begins to look the same (Noss 1983). Diversification can all too easily become homogenization.

One way to look at the relationship between biodiversity and sustainability is that biodiversity provides enduring options for sustainable management. As Leopold (1949) noted, "to keep every cog and wheel is the first precaution of intelligent tinkering." Particularly when the future is unpredictable, as it surely is now with human population growth, irrational energy policies, climate change, ozone depletion, and other global problems, it makes sense to maintain as many options as possible. A landscape with a great diversity of habitats, species, and genotypes is likely to be more adaptable to change than is a monoculture.

A most important question we must ask with regard to sustainability is What do we wish to sustain and why? This is essentially an issue of goal setting. We need to pay more attention to where we wish to head with our land management programs and to what values are behind those objectives. If our goal is only to maintain an approximately even flow of wood products, then we have a seemingly easier task than if we have to worry about sustaining the food webs and nutrient cycles that maintain soil productivity. Of course, in the long run we must think about maintaining soils and ecological processes if we want a sustained yield of wood products. Maintaining watershed integrity may require even more restraint.

The National Forest Management Act (NFMA) of 1976 has sustainability provisions. It states that timber is to be cut only where "soil, slope, or other watershed conditions will not be irrevocably damaged and where land can, with assurance, be restocked within five years after logging." The regulations implementing the NFMA require the Forest Service to manage the land so as not to impair its multiple-use productivity and so as to consider all renewable resources (Norse et al. 1986). Although the NFMA guidelines appeal more to an agricultural than to an ecological criterion for sustainability, ecological thinking does underlie the NFMA, more so than for virtually any other law on the books. Despite all of the rhetoric about multiple use, however, the overriding goal driving the management of national forests still seems to be getting the cut out.

The ecosystem argument for sustainability is good as far as it goes, but it does not guarantee that native biodiversity will persist. Ecological arguments can suggest many different things, depending on the goal. We may be able to design a forest that provides wood products in perpetuity, soil and watershed integrity, and persistence of most native species but fails to maintain highly sensitive species, such as grizzly bears or Florida panthers, or suitable conditions for continued evolution of species. And there are still other qualities of forests that are seldom considered in forestry or ecology but yet are important to us in immeasurable ways. These other values include wildness and naturalness. A perfectly functioning tree farm, even if quite diverse in native species and genotypes, is not good enough; it is too much like a machine.

Sustainability, then, is a multifaceted and relative concept. It must be defined in terms of (1) time period (are we sustaining forests for two, three, or four rotations or for millennia?) and (2) proportion of ecosystem structure, function, and composition maintained (Franklin et al. 1981). Structure includes dead wood that is not salvaged, function includes wildfires and floods that are sometimes inconvenient for humans, and composition includes species, such as large carnivores, that do not like to have us around and others, such as soil invertebrates and fungi, that most people find very boring. Managed forests should be compared in terms of how well they maintain all of their native components over time, not just those that are convenient for human society.

Exploited forests must be measured against some kind of standard to assess their relative sustainability. The best control area against which exploited forests can be compared is the natural, unexploited forest, as was recognized over fifty years ago by Leopold (1941), who stated that wilderness provides a "base-datum of normality, a picture of how healthy land maintains itself as an organism." Unexploited forests also provide qualities of wildness and naturalness that no exploited forest can match. But we can compare exploited forests in terms of these seemingly intangible qualities as well. Anderson (1991) offered three criteria for assessing the relative naturalness of any area: (1) the amount of cultural energy required to maintain the system in its present state, (2) the extent to which the system would change if humans were removed from the scene, and (3) the proportion of the fauna and flora composed of native versus nonnative species. Forests that can take care of themselves, would change little if we left them alone, and are made up of native species are more natural—and, I submit, more sustainable—than are humanized forests. So naturalness and sustainability may not be so far apart as conservation criteria.

We tend to think of sustainable forestry as sustainable exploitation. But this is an arrogant view and clearly not the whole picture. Acknowledging that forests must and should be used by society for a variety of products and services, I believe that it is also essential that large areas be left essentially unmanaged except for the protective and restorative management now needed for all but the most remote regions on Earth. A sustainable forest, at a regional scale, contains unexploited as well as exploited areas. In our rush to experiment with the latest techniques in New Forestry, I fear we are forgetting about the need for control areas, without which our experiments are meaningless. Without a healthy dose of humility and restraint, the sustainability concept can be dangerous.

FOREST MANAGEMENT

Forest management includes a wide range of practices, from intensive tree farming to leaving areas alone. Some people find the whole concept of management insolent and pernicious, but I am afraid that we are stuck with it in the short term and with five and a half billion people on Earth. In any case, management and stewardship are at least potentially a big step above earlier practices of cut-and-run exploitation.

The effects of forest management on biodiversity differ widely among practices and circumstances. Management may be (1) negative, threatening biodiversity by failing to maintain or mimic natural ecosystem processes and take into consideration the needs of sensitive species; (2) neutral, if it emulates nature and substitutes for natural disturbance-recovery cycles; or (3) positive, if it is oriented toward restoring forests abused by past management practices. Most of us will agree that negative forest management must be phased out as quickly as possible, everywhere, and be replaced by management that is at least neutral with respect to biodiversity and, wherever possible, positive, being concerned with healing and nurturing abused forests.

IDENTIFYING TRAJECTORIES OF IMPOVERISHMENT

How are we to set objectives for protecting biodiversity in managed forests and for healing abused forests? A first step is to identify the trends or trajectories of landscape change that have been associated with biotic impoverishment in a region. Biotic impoverishment in forest landscapes can be linked to habitat changes that exceed the adaptive capacities of at least some native species. These habitat changes represent a shift in landscape conditions to a present condition that is less favorable to native biodiversity than the natural or historic condition of the landscape. Although natural condition is notoriously difficult to define and is always changing, we do not have to quantify natural condition to identify the ways in which the present landscape has departed from it and how those changes threaten native biodiversity.

The shift from natural conditions affects several characteristics of the forest landscape, notably stand age distribution, structural diversity, fragmentation, fire regimes, road density, and rare species. The pathway of change in each of these characteristics can be thought of as a trajectory from the natural to the managed conditions, and the landscape as a whole is transformed along several trajectories simultaneously (e.g., from older to younger stands, from complex to simple stands, and from contiguous to isolated patches). These trajectories are often intercorrelated but can vary independently of one another and proceed at different rates. There will be regional and local differences in the strength of each trend, and many landscapes will demonstrate only a subset of these trends. In some landscapes, other trends, such as livestock grazing or air pollution, may be more important than the seven described below.

Younger Forests

Many unexploited forests are dominated by older age classes. An individual landscape may shift radically in age class distributions over time because of stand-replacing disturbances and subsequent succession; therefore, age class (seral stage) distributions are best estimated at a regional scale. In the Pacific Northwest, it is estimated that some 60 to 70 percent of commercial forest land in the Douglas fir region in the early 1800s was old growth (Franklin and Spies 1984). Old growth was also the most abundant seral stage of the eastern deciduous forests (Greller 1988) and the longleaf pine forests that dominated the southeastern coastal plain (Platt, Evans, and Rathbun 1988), among others.

With logging, forest age classes across these regions shifted to favor younger stages. No more than 13 percent of the presettlement old growth remains in the Pacific Northwest (Norse 1990); the figure is generally smaller in other regions of the United States, excluding Alaska. With short rotations for wood fiber production, dominance by younger age classes persists indefinitely. The consequences of this shift are most severe for those species closely associated with old growth, such as the northern spotted owl and a number of other species in the Northwest (Carey 1989) and the red-cockaded woodpecker in the Southeast (Jackson 1971).

Simplified Stands

With conversion of natural forests to plantations, stand structure is greatly simplified. Traditional clear-cutting and site preparation practices remove most of the coarse woody debris before planting. The barren site that remains differs dramatically from a stand affected by natural disturbance such as fire or windthrow. In Douglas fir forests of the Northwest, research has demonstrated that snags, downed logs, and surviving trees persist through much of the successional sequence after a natural disturbance and provide sites for plant regeneration, water retention, and nutrient cycling, as well as wildlife habitat (Franklin et al. 1981; Maser et al. 1988).

Douglas fir plantations with reduced structural diversity have been shown to contain reduced species richness of breeding birds, small mammals, and amphibians and greatly reduced abundances of birds and amphibians relative to natural forests of any age in Oregon (Hansen et al. 1991). In Florida, species richness and density of breeding birds decline when natural longleaf pine forests are converted to simplified plantations (Repenning and Labisky 1985).

Smaller Patches

As a natural forest is converted to an intensively managed forest, the most pronounced change at the landscape scale is often habitat fragmentation (Harris 1984; Franklin and Forman 1987; Hunter 1990). In the later stages of fragmentation, remnant patches of natural forest may be quite small. The classic documentation of forest fragmentation in Cadiz Township, Wisconsin, showed a reduction in average forest patch size from 8,724 hectares in 1831, to 36.9 hectares in 1882, 13.8 hectares in 1902, 7.4 hectares in 1935, and 5.8 hectares in 1950 (Curtis 1956). Agriculture was responsible for the fragmentation in this case, but a similar (although often less severe) process occurs with even-age management.

As forest patch size declines, the amount of forest edge per unit forest area increases. Wildlife managers traditionally have favored forest harvest patterns that intersperse forage and cover areas because most game animals are edge adapted. Recent research, however, has documented many deleterious consequences of edge for sensitive forest interior species (Noss 1983; Wilcove, McLellan, and Dobson 1986).

The trend toward decreasing patch size and increasing edge is occurring in many regions subjected to intensive logging today. Between 1972 and 1987, for example, average forest patch size in two ranger districts of the Willamette National Forest, Oregon, decreased by 17 percent, the amount of forest-clear-cut edge doubled, and the amount of forest interior at least 100 meters from edge declined by 18 percent (Ripple, Bradshaw, and Spies 1991). In the Olympic National Forest, Washington, more than 87 percent of the old growth in 1940 was in patches larger than 4,000 hectares; in 1988, only one patch larger than 4,000 hectares remained, and 60 percent of the old growth was in patches smaller than 40 hectares. Forty-one percent of the remaining old growth in 1988 was within 170 meters of edge and therefore vulnerable to blowdown and other edge effects (Morrison 1990).

Isolated Patches

The other critical variable in habitat fragmentation, besides patch size, is isolation. As forest fragmentation proceeds, patches of remaining natural forest become more isolated from one another. Dispersal is the key factor determining the regional effects of fragmentation. Movement of individuals between patches must be great enough to balance extirpation from local patches for a species to persist in a fragmented landscape (Wiens 1989). Some species, especially those of early successional habitats, are adept at dispersing between isolated patches of suitable habitat. Species of late successional habitats, such as flightless invertebrates found in old-growth forests, are usually poorer dispersers and are more vulnerable to the isolating effects of fragmentation (Horn 1978; den Boer 1990; Moldenke and Lattin 1990).

In general, dispersal is easier through habitats similar in structure to the habitat in which a species lives (Wiens 1989). For probably most late-successional species, patches of old growth distributed in a matrix of tree plantations of various ages are functionally less isolated than forest patches surrounded by agricultural fields or fresh clear-cuts. Harris (1984) suggested that old growth patches in the Pacific Northwest should be surrounded by a series of long-rotation stands harvested on different schedules.

Fewer Fires

Many forest types and species' life histories have been shaped evolutionarily by fire (Mutch 1970). Human activities may either increase or decrease fire frequency and intensity, with marked consequences for biodiversity. In the New Jersey pine barrens, urbanization has generally lengthened fire return intervals (Forman and Boerner 1981). The number of fires per year (regional fire frequency) has actually increased, but average fire size has decreased dramatically, reducing point fire frequencies. In the southeastern coastal plain, fire suppression has been largely responsible for the invasion of natural, open-structured longleaf pine forests by dense hardwoods (Myers 1985). Longleaf pine savannas protected from fire may be less rich floristically than their well-burned counterparts (Walker and Peet 1983). As a dense understory develops, longleaf pine stands become unsuitable for red-cockaded woodpeckers (Jackson 1971).

Similarly, ponderosa pine forests of the West have been greatly altered by fire suppression. In central Oregon, early foresters reported 10 to 30 trees per acre with average diameters of over 17 inches. Today, the norm is 300 trees per acre and average diameters of less than 10 inches (Daniel 1990). Increased density of trees due to fire suppression may promote insect infestations, disease, and vulnerability to catastrophic fire (Anderson, Carlson, and Wakimoto 1987; Habeck 1990).

The trend toward fewer fires has often been irregular. Whittaker (1960) documented an early historical period in the Siskiyou Mountains of Oregon and California when fire frequency increased markedly with white settlement, followed by a more recent period of protection from fire. A similar trend reversal has occurred in other regions, such as the New Jersey pine barrens (Forman and Boerner 1981). However, drought cycles and other climatic changes on a scale of decades to centuries may be responsible for many observed patterns in fire history (Clark 1990).

More Roads

Forestry activities usually depend on access by roads. In the Northwest, each square mile of intensively managed commercial forest land requires an average of 5 miles of road (Norse 1990). Roads directly eliminate natural habitat, restrict movements of small animals that hesitate to cross even narrow swaths (Mader 1984), create edge effects and exacerbate blowdown (Franklin and Forman 1987), contribute massive amounts of sediments to streams in steep terrain (Swanson, Franklin, and Sedell 1990), lead to roadkills (Harris and Gallagher 1989), encourage invasions of forest pests and diseases (Schowalter 1988), and provide access to legal and illegal hunters, reducing habitat effectiveness for many large mammals (Lyon 1983; Thiel 1985). For animals vulnerable to human activity, roads and their associated influences represent some of the greatest threats associated with forest management.

More Endangered Species

Although the proximate causes of species endangerment are diverse, the trends in forest condition identified above and observed in many landscapes are unfavorable to the survival of many sensitive species. The new landscape created by human activities may be very different from what native species have experienced during their evolutionary histories. We may be just beginning to see the biotic consequences of landscape change in much of North America, as there is often a time lag between habitat changes and responses of the biota at population and community levels. Site fidelity may cause many birds, for example, to remain in or return to an area long after it has become unsuitable for reproduction (Wiens 1989). Small populations of animals (and, eventually, of plants) restricted to forest fragments and isolated by roads and other barriers may not be viable in the long term. The 750 species of plants and animals federally listed as endangered or threatened in the United States, plus thousands of candidate species, are only the tip of the iceberg of biotic impoverishment.

REVERSING TRAJECTORIES OF IMPOVERISHMENT

Is it possible to devise a forest management regime that can reverse these trajectories of impoverishment, allow biodiversity to recover, and thus sustain forests as well as forestry into the future? The goal of reversing the kinds of landscape changes just described does not imply that returning to some pristine natural condition is either possible or desirable in most cases. The presettlement forest in a particular landscape was one frame in a very long movie, played on a projector with no reverse switch. Native human cultures, as well as climate, physical habitat, and accidents of ecological history, have long shaped forests. In some cases, the use of forests by native cultures was "sustainable" by the criteria we are discussing here, and in other cases it surely was not (Noss 1985). The presettlement forest was only one of many possible communities that might have developed in a particular landscape after the last major disturbance, but we have little idea of what these alternative steady states might be.

Reversing trajectories of impoverishment is restoration, surely enough, but it is not returning; it is becoming. It is becoming something more secure for sensitive native species, more natural, and more sustainable by any reasonable criteria. The restored, sustainable forest will not be the same as the pre-European settlement forest, but it will be closer to it in some fundamental ways than the present, exploited forest and, therefore, more similar to the ecological theater in which the native species of a region evolved. Below I suggest a few things that can be done to allow forests to recover to a sustainable condition.

Older Forests

Natural forests, as discussed above, are often dominated by old growth, and species associated with older forests tend to be more vulnerable to fragmentation than are early successional species. For a region that has lost much of its old-growth forest, protecting every scrap of old growth that remains and allowing much of the younger forest landscape to return to an old-growth condition may be essential for achieving true sustainability.

Federal forest managers, in some cases, are considering longer harvest rotations for planted forests. The committee of scientists appointed by the House Agriculture Committee to develop options for forest management and protection in the Pacific Northwest, commonly known as the "Gang of

Four," recommended rotations of 180 years or more in some alternatives, particularly to protect watersheds with important fishery values. But old growth in this region and elsewhere continues to be cut—it is *not* being sustained—so it seems that the values of old growth are still not fully appreciated by the forestry establishment or by Congress.

Managing a forest landscape toward increasingly older age classes will be economically costly in the short term but can be expected to help maintain soil fertility and enhance populations of species associated with older seral stages, many of which are threatened with extinction under current management.

Structurally Complex Stands

A number of silvicultural manipulations have been suggested for the purpose of retaining or restoring structural diversity in managed stands. Snag retention in clear-cuts has long been used to benefit cavity-nesting birds and mammals (Thomas et al. 1979). Leaving "dead and down" woody material in harvest units also has demonstrable wildlife benefits (Maser et al. 1979). The increasing emphasis on maintaining habitat structure in managed stands falls under the banner of the New Forestry of Jerry Franklin and colleagues in the Pacific Northwest (Franklin 1989; Gillis 1990).

Retention of coarse woody debris and mature green trees, scattered or in clumps throughout a harvest unit, is a prominent feature of New Forestry. Unlike a shelterwood cut, partial retention harvests leave green trees (about 30 percent of those before cutting) at least through the subsequent rotation. A site harvested under New Forestry might resemble a site after a natural fire or windstorm as much as, or more than, it resembles a clear-cut. Retention cuts are expected (but not proven) to mitigate the effects of harvest on soil and soil organisms and to provide habitat for many species. Sloppy harvest areas also may be less of a barrier than traditional clear-cuts to dispersing animals that require cover. New Forestry measures may prove effective in restoring structural and species diversity and hastening the development of old-growth characteristics in plantations. In my view, however, neither New Forestry nor any other method of harvest is appropriate for our few remaining natural forests, old growth or otherwise, although thinning may help restore fire-suppressed stands.

Larger Patches

Particularly on federal lands, the most common cutting pattern in recent years has been a "staggered setting" or a dispersed patch pattern that distributes clear-cuts throughout a landscape. At its extreme, this pattern is a checkerboard of clear-cuts and remaining forest that maximizes fragmentation (Franklin and Forman 1987). Although, as noted above, wildlife managers have traditionally favored this kind of harvest scheduling because it benefits edge-adapted game species, the negative side of edge effects has recently become clear (Noss 1983; Wilcove, McLellan, and Dobson 1986). Alternative designs aggregate cutting units and thereby retain larger patches of intact forest and connecting corridors in the landscape. Interior forest species sensitive to edge effects would be expected to survive better under these conditions.

Some of the more recent national forest plans contain fragmentation reduction standards and guidelines. For example, the forest plan for the Mt. Hood National Forest in Oregon specifies that "harvest unit selection should favor existing isolated, relatively small blocks of forest" and that "harvest units should be located minimizing fragmentation of large blocks of old growth by placing the harvest units on the margin of the large blocks" (Mt. Hood National Forest Plan IV:67). Of course, in landscapes already heavily fragmented, no further cutting may be warranted until young stands have matured and filled in some of the holes in the forest matrix. In such landscapes, restoration consists mostly of "growing habitat fragments" (Janzen 1988).

Connected Patches

Connectivity for dispersal is a problem most intense for late-successional forest species. For such species, blocks of habitat that are close together are better than blocks far apart, and blocks of habitat connected by suitable habitat linkages are preferable to those that are isolated (Harris 1984; Noss 1987; Thomas et al. 1990). Habitat linkages may be either distinct forest corridors or a structurally rich landscape matrix that permits safe movement of target species between forest patches. We have much to learn, however, about what constitutes movement barriers and corridors for various species.

The Interagency Scientific Committee (ISC) on the northern spotted owl noted that spotted owls disperse in random directions and would not be expected to use linear corridors (Thomas et al. 1990). Instead, the ISC recommended that 50 percent of the landscape matrix outside proposed spotted owl habitat conservation areas be maintained in forest averaging at least 11 inches dbh (diameter at breast height) and 40 percent canopy closure. Other species that shun openings, lack wings, or are more vulnerable to human activity may require intact forest corridors of some minimum width (Noss 1993). Roads intersecting corridors can be a serious problem because some species rarely cross roads or risk their lives when doing so. So that the effects of roads on wildlife movements can be mitigated, bridges or wildlife underpasses should be constructed where feasible (Harris and Gallagher 1989; Noss 1993).

Prescribed Fires

Forest landscapes, if sufficiently large and wild, might incorporate a natural disturbance regime (Pickett and Thompson 1978), and lightning fires could be allowed to burn uncontrolled. Relying on a natural disturbance regime to maintain forest diversity is generally impractical in managed forests or small protected areas, however. Fire-dependent communities, such as ponderosa pine or longleaf pine, must usually be maintained through prescribed burning, sometimes preceded or accompanied by thinning of small woody stems to reduce fuel levels. Wildfires in fire-suppressed stands can be catastrophic (Habeck 1990).

Prescribed fire and accompanying silvicultural techniques to reduce wildfire potential are rather commonly applied in the West (Martin, Kauffman, and Landsburg 1989). In managed pinelands in the Southeast, prescribed fire is often used to prepare sites for pine regeneration, to control competing vegetation, to reduce the potential for hazardous fires, and to control disease and insect pests (Abrahamson and Hartnett 1990). Management burns, however, are usually conducted during a season (winter or early spring) different from when natural fires occur (usually late spring and summer) and have less than optimal effects from the standpoint of native biodiversity (Myers 1990). Furthermore, although prescribed burning is relatively simple in open-canopied forests, such as longleaf pine or ponderosa pine, applying fire to forests that naturally experience high-intensity fires, such as lodgepole pine or sand pine, is another story. The ecological effects of excluding fire from such forests and clear-cutting them instead are not well known (Myers 1990). A remaining problem in fire management is public concern about air quality, although public education about the ecological benefits of fire would be helpful in alleviating this problem.

Lower Road Density

In many landscapes, perhaps the single most effective restoration measure for sensitive wildlife would be road closures. The sensitivity of hunted elk populations to road density, for example, is well established. An open road density of 1 mile per square mile of habitat can decrease habitat effectiveness for elk by 40 percent compared to roadless watersheds; as road density increases to 6 miles of road per square mile, elk habitat use falls to zero (Lyon 1983; Wisdom et al. 1986). Wolves are even more sensitive to the access that roads provide to legal and illegal hunters and may not be able to maintain populations where road density exceeds 0.9 mile per square mile (Thiel 1985; Mech et al. 1988). Closure of roads to public use can benefit these and many other species. However, the reduction of movement barrier effects on small animals, the inhibition of landslides, and the limitation of sediment inputs to streams require that roads be ripped and revegetated (Noss 1991).

Recovering Species

If habitat protection and restoration practices such as those described above are implemented, species that have declined because of intensive forest management may at least show a reduced rate of decline. With time, many populations can be expected to stabilize and eventually recover to a higher density or wider distribution. This pattern of a reduced rate of decline followed by stabilization and eventual recovery is the predicted outcome of implementing the ISC conservation strategy for the northern spotted owl (Thomas et al. 1990). We can predict that many other imperiled species would show similar recovery with a shift to older forests, complex stands, larger patches, connected patches, prescribed fires that mimic natural fire regimes, and reduced road density. In the case of the spotted owl, timber will still be harvested in the region throughout the recovery process, although certainly at a reduced level.

To restore entire ecosystems to a sustainable condition, people must make short-term economic adjustments. But these adjustments need not result in fewer jobs in forestry-related activities. Restoration forestry—

including the thinning of plantations, the replanting of clear-cuts and thinned plantations with a diversity of native species, the closing and revegetating of roads, and the rehabilitation of streams—is labor-intensive. Restoration can keep many workers in the woods without expensive job retraining programs. Adjustments in budgets, both within and among agencies, will be needed to employ restoration workers. Financial incentives for restoration on private forest lands are also necessary, as are government programs to encourage conservation and recycling of wood products, thereby reducing the demand for raw logs.

What is really needed to restore America's forests is probably no less than a new national Civilian Conservation Corps (CCC), with projects based on modern, scientifically informed conservation priorities. The old CCC, as Leopold (1949) lamented, was too concerned with building roads into lovely country and sometimes did more harm than good to wildlands. The new CCC must be guided by a land ethic and by what has been learned in ecology over the six decades since the old CCC was in operation. The payoff of such a program for society will be a healthier, more livable environment for future generations of humans and nonhumans alike and a reduced need for costly crisis management, such as endangered species programs.

MONITORING AND ASSESSMENT

Every move toward sustainable forestry and sustainable forests is an experiment. We develop hypotheses about what is wrong with current management and about what must be changed to make forestry and forests more sustainable. We conduct experiments to see if our new approaches work any better than did past approaches. As Walters and Holling (1990) noted, "every major change in harvesting rates and management policies is in fact a perturbation experiment with highly uncertain outcome, no matter how skillful the management agency is in marshalling evidence and arguments in support of the change."

Because we are engaged in applied science and are more concerned with problem solving than with knowledge for its own sake, our hypotheses must be relevant to real-world problems. Moreover, our hypotheses and experiments must be guided by goals and objectives that indicate how we want the world to be. Suitable goals for biodiversity conservation include maintaining viable populations of native species in natural patterns of distribution and abundance while allowing natural ecological and evolutionary change.

INDICATORS

Whatever the management goals, progress toward those goals must be tracked. To track progress, one must be able to assess present environmental conditions in quantifiable terms and monitor changes in those conditions over time. Monitoring and assessment depend on indicators, which may be defined as measurable attributes of species, habitats, or processes that correspond to environmental endpoints of concern (Noss 1990). Biodiversity in the broad sense is an environmental endpoint of increasing interest to the public. More specifically, endpoints include species richness, the health and abundance of special-interest species (including game species and rare species), a pollution-free environment, and aesthetic beauty.

If historical information in the form of a time series is available, indicators can be used to analyze past trends in landscape condition and biodiversity. That is, they can be used to measure departures from natural conditions since settlement or logging began, as discussed above. For instance, forest pattern and patch size indicators, manipulated by aid of Geographic Information Systems (GIS), were used by Morrison (1990) and Ripple, Bradshaw, and Spies (1991) to measure the extent of forest fragmentation in their Washington and Oregon study areas, respectively.

Elsewhere I presented a framework for monitoring biodiversity by use of indicators of ecosystem composition, structure, and function at several levels of biological organization (Noss 1990). Table 1.1 lists some stand- and landscape-level indicators that may be useful for monitoring trends in forest condition. No manager would want to use this entire "laundry list" of indicators. Rather, indicators should be selected to correspond to the ecosystems, landscapes, species, and management goals for that particular situation.

As an example, suppose that our landscape is in northern Florida. There are remnant "islands" of mature longleaf pine flatwoods and sandhills in a matrix of slash pine plantations, none older than thirty years. Wetlands, especially depression marshes, are scattered throughout the landscape, and hardwood forests are on steep slopes and in the bottomlands along rivers. Management objectives include restoring the natural open-canopy

TABLE 1.1. **Indicators of Forest Landscape Condition**

Forest Age

    Frequency distribution of seral stages (age classes) for each forest type and across all types

    Average and range of tree ages within defined seral stages

Forest Structure

    Ratio of area of natural forest of all ages to area in clear-cuts and plantations

    Abundance and density of snags, downed logs, and other defined structural elements in various size and decay classes

    Spatial dispersion of structural elements and patches

    Foliage density and layering (profiles) and horizontal diversity of foliage profiles in stand

    Canopy density and size and dispersion of canopy openings

    Diversity of tree ages or sizes in stand

Patch Size (and Related Variables)

    Patch size frequency distribution for each seral stage and forest type and across all stages and types

    Patch size diversity index

    Size frequency distribution of late-successional interior forest patches (minus defined edge zone, usually 100–200 m)

    Total amount of late-successional forest interior habitat

    Total amount of forest patch perimeter and edge zone

    Forest patch perimeter:area ratio

    Edge zone:interior zone ratio

    Fractal dimension

    Patch shape indices

    Patch density

    Fragmentation indices

Patch Isolation (and Related Variables)

    Interpatch distance (mean, median, range) for all forest patches and for late-successional forest patches

TABLE 1.1. **Indicators of Forest Landscape Condition** (*continued*)

Juxtaposition measures (percentage of area within a defined distance from patch occupied by different habitat types, length of patch border adjacent to different habitat types)

Structural contrast (magnitude of difference between adjacent habitats, measured for various structural attributes)

Fire Regime

Frequency, return interval, or rotation period

Areal extent

Intensity or severity

Seasonality or periodicity

Predictability or variability

Roads

Road density (mi/mi² or km/km²) for different classes of road and all road classes combined

Percentage of forest in roadless area (5,000 ac or larger, and other RARE II criteria) outside designated wilderness

Percentage and area of existing roadless area retained at end of each decade

Miles of roads constructed, reconstructed, and closed (seasonally and permanently) each decade (and ratios)

Amount of roadless area restored through permanent road closures and revegetation

Sensitive Species

Demographic parameters (abundance, density, cover or importance value, fertility, recruitment rate, survivorship, mortality rate, etc.)

Genetic and health parameters (allelic diversity, heterozygosity, individual growth rate, fecundity, body mass, stress hormone levels, etc.)

---

structure and diverse groundcover of longleaf pine stands (many of which have been invaded by hardwoods because of the lack of fire); thinning plantations to mimic the natural pineland structure in corridors connecting remnant longleaf stands (to encourage the dispersal of red-cockaded woodpeckers and other pineland fauna); maintaining natural upland-wetland gradients and seepage slope bogs, which are especially rich in endemic plant species; maintaining intact riparian corridors; and reducing the poaching of gopher tortoises and black bears.

Key indicators to assess current conditions and monitor the restoration process in this example would include the density, age distribution, and spatial dispersion of trees within stands; stand size and shape indices; connectivity of stands across the landscape; fire frequency and seasonality for each vegetation type; groundcover richness; juxtaposition measures; riparian corridor width and structure; road density (which corresponds to access for poachers); and demographic processes (e.g., recruitment and survivorship) of sensitive indicator species.

ADAPTIVE MANAGEMENT

Monitoring and assessment programs, placed in a hypothesis-testing framework and using quantifiable indicators, are essential for measuring progress toward goals and comparing the long-term effects of alternative management practices. But how do we apply what we learn from these experiments? Management experiments must function as an adaptive process (Walters 1986).

The present condition of biodiversity in many forest landscapes suggests that past management approaches, in general, have been less than sustainable. We have learned enough about forest ecology in recent years to know that some practices (e.g., retaining woody debris in harvest units and maintaining landscape-scale connectivity) are more likely to protect biodiversity than are their status-quo alternatives. But nature is quite capable of surprising us. New approaches must be tested experimentally and results monitored. Management policy must be based on the best current knowledge and be flexible enough to respond quickly to new information.

Adaptive management (Holling 1978, Walters 1986) is simple in concept—it is learning from experience—but can be difficult in operation. The trick is to make management responsive but not capricious or erratic. Obviously, it would be foolhardy to change management direction on the basis of every new study. Rather, scientific advisory panels should serve the purpose of reviewing available data, forming a consensus about appropriate land allocations and management practices, and providing counsel on these matters to policy makers. The basic goals of biodiversity protection and restoration could remain essentially the same and would provide consistency and direction to management plans; only the details of the plans would change as knowledge is improved.

Relying on a consensus of experts to develop management policy and practices, monitoring, and adjusting the program as suggested by the results of monitoring constitute passive adaptive management. The alternative and preferred approach of active adaptive management requires a blend of research and monitoring; alternative management regimes are tested as experiments (Walters 1986, Thomas et al. 1990). In either case, scientific advisory panels could play an important role in interpreting results to decision makers. It is essential that an advisory panel represent the best available expertise from a variety of relevant disciplines and that their reports not gather dust on the shelves of bureaucrats.

SUSTAINING FORESTS IN THE FACE OF UNCERTAINTY

The experimental, monitoring-intensive, adaptive management approach to sustainable forestry I outlined above is built on the premise that we can learn how to manage forests much better than we have in the past. I now want to add a small caveat: We really do not know what we are doing. Although the experimental, adaptive approach is designed to provide more certainty about how forests might be managed sustainably, uncertainty will remain the dominant theme in day-to-day decisions about which lands to log and which to place under strict protection, which harvest and regeneration methods to use in unit A and unit B, and how much wood and other products a region can produce over a period of years without risking degradation of ecosystem functions and loss of native biodiversity. We are learning more all the time, but nature will always remain several steps ahead of us. As the ecologist Frank Egler once remarked, "nature is not only more complex than we think, but more complex than we can ever think."

How can a proper humility and caution be inserted into environmental policy making at all levels? A good start would be for scientists themselves to take more responsibility for their recommendations—a responsibility not only to practice good, credible science, but also to ensure that their science truly contributes to biodiversity conservation. Again, applied science is problem solving, and our problem is the biodiversity crisis. Policy makers and the pubic at large are looking more and more to scientists for advice on how to solve environmental problems, such as how to sustain forests. Scientific recommendations will continue to be derided or ignored by congressmen with special interests, but generally they will carry considerable weight. This newfound influence in the policy arena places a tremendous ethical burden on ecologists.

Both scientists and the public often assume that the relationship between science and environmental policy is direct—like a straight arrow. Scientists supply objective facts, without much interpretation and without any recommendations, and policy makers make informed decisions. If a scientist tries to put these facts in a broader context or suggest that certain policies be changed, he or she may be accused of going beyond the bounds of science.

In reality, the relationship between science and policy is more diffuse. Facts do not speak for themselves; they must be interpreted within some broader context. But interpretation means that values enter in—values of the scientists, of the policy makers, and of the public. Value-free science is a myth, an impossibility (Longino 1990). It is not at all inappropriate for scientists to play a direct role in the policy-making process. In my view, environmental policy is too important to be left to the policy makers, most of whom know little and care little about forests or sustainability in the broad sense.

Perhaps the most useful model for how science can positively influence policy is one in which scientists present various options for resource management to the decision makers. A good example is the recent Gang of Four process mentioned above, where four forest scientists were asked by the House Agriculture Committee to come up with options for managing old-growth forests in the Pacific Northwest. After identifying options, scientists may then conduct a "risk analysis" (which may amount to nothing more than their collective expert opinion) to evaluate the potential costs and benefits of each management alternative in terms of social, economic, and environmental values.

There is a difference of opinion on how options should be presented to policy makers, however. One course is to present them without comment on preference. This was the approach followed by the Gang of Four. Another course is to select a preferred alternative, the standard procedure for national forest plans and all other federal actions subject to the National Environmental Policy Act (NEPA), as the ISC did for the spotted owl. Or, leaving it open a bit more, scientists might select a "critical threshold" below which all options are biologically unacceptable. This is the course I would prefer. We cannot count on politicians to make biologically prudent decisions without rather explicit guidance.

Ecological investigations, like all of science, may lead to "the truth," but there is considerable uncertainty. We risk two major types of errors, usually called type I and type II, with potentially serious consequences. Generally, scientists worry most about type I error, of rejecting a true null hypothesis and claiming an effect when none exists. The concept of statistical significance is based on minimizing the probability of type I error. The lower the risk of type I error, however, the higher the risk of type II error. In applied ecology, type II errors can be more dangerous than type I errors.

If we commit a type I error (e.g., by saying that present levels of logging in the Northwest will drive the spotted owl to extinction in fifty years but then that turns out not to be the case), the policy response may have been to protect more land than was necessary. We erred on the side of preservation. Consequently, we could lose some credibility as scientists and we could lose political support. These are legitimate concerns. There is already political backlash about spotted owl protection, with Senator Mark Hatfield saying that biology has "run amok." However, a historical analysis shows that the credibility of science has continued to increase in the public eye despite numerous false predictions (K. S. Shrader-Frechette and E. D. McCoy, Method in community ecology: Applied biology and the logic of case studies, in preparation).

If scientists instead commit a type II error and say that logging poses no danger to spotted owls (the predictable position of the timber industry), then society may protect less than necessary and err on the side of development or industry. In general, type II errors of predicting no significant effect of various human actions, such as logging, are bound to result in the loss of natural areas. The cumulative impact may well be extinction of species.

Ideally, we strive to avoid both types of errors and seek the truth, however elusive. But we have an ethical obligation to favor biodiversity over industry in the face of uncertainty. In practice, erring on the side of preservation—the prudent and conservative course—means minimizing the influence of human activities on the land. It means experimenting cautiously with new approaches to forestry and being properly skeptical about claims for sustainability. It means drastically reducing our demand for wood products, through conservation, re-use, recycling, and human population control, so that the greatest possible amount of natural forest can be left wild and degraded forest lands have time to be restored to health. The prudent approach to sustainability puts forests above forestry.

ACKNOWLEDGMENT

This is contribution 626 from the University of Idaho, Forestry and Wildlife Resources Experiment Station.

REFERENCES

Abrahamson, W. G., and D. C. Hartnett. 1990. Pine flatwoods and dry prairies. In *Ecosystems of Florida*, eds. R. L. Myers and J. J. Ewel, 103-149. Orlando, FL: University of Central Florida Press.

Anderson, J. E. 1991. A conceptual framework for evaluating and quantifying naturalness. *Conservation Biology* 5:347-352.

Anderson, L., C. E. Carlson, and R. H. Wakimoto. 1987. Forest fire frequency and western spruce budworm outbreaks in western Montana. *Forest Ecology and Management* 22:251-260.

Carey, A. B. 1989. Wildlife associated with old-growth forests in the Pacific Northwest. *Natural Areas Journal* 9:151-162.

Clark, J. S. 1990. Fire and climate change during the last 750 yr in northwestern Minnesota. *Ecological Monographs* 60:135-159.

Curtis, J. T. 1956. The modification of mid-latitude grasslands and forests by man. In *Man's role in changing the face of the earth*, ed. W. L. Thomas, 721-736. Chicago: University of Chicago Press.

Daniel, J. 1990. Old growth on the dry side. *High Country News* 22(22):27-28.

den Boer, P. J. 1990. The survival value of dispersal in terrestrial arthropods. *Biological Conservation* 54:175-192.

Forman, R. T. T., and R. E. Boerner. 1981. Fire frequency and the pine barrens of New Jersey. *Bulletin of the Torrey Botanical Club* 108:34-50.

Franklin, J. F. 1989. Toward a new forestry. *American Forests* Nov./Dec.:37-44.

Franklin, J. F., K. Cromack, W. Denison, A. McKee, C. Maser, J. Sedell, F. Swanson, and G. Juday. 1981. Ecological characteristics of old-growth Douglas-fir forests. Gen. Tech. Rep. PNW-118. Portland, OR: USDA Forest Service, Pacific Northwest Forest and Range Experiment Station.

Franklin, J. F., and R. T. T. Forman. 1987. Creating landscape patterns by cutting: Ecological consequences and principles. *Landscape Ecology* 1:5-18.

Franklin, J. F., and T. A. Spies. 1984. Characteristics of old-growth Douglas-fir forests. In *New forests for a changing world*, 328-334. Bethesda, MD: Society of American Foresters.

Gillis, A. M. 1990. The new forestry. *BioScience* 40:558-562.

Greller, A. M. 1988. Deciduous forest. In *North American terrestrial vegetation*, eds. M. G. Barbour and W. D. Billings, 288-316. Cambridge: Cambridge University Press.

Habeck, J. R. 1990. Old-growth ponderosa pine-western larch forests in western Montana: Ecology and management. *Northwest Environmental Journal* 6:271-292.

Hansen, A. J., T. A. Spies, F. J. Swanson, and J. L. Ohmann. 1991. Conserving biodiversity in managed forests: Lessons from natural forests. *BioScience* 41:382-392.

Harris, L. D. 1984. *The fragmented forest: Island biogeography theory and the preservation of biotic diversity*, 211. Chicago: University of Chicago Press.

Harris, L. D., and P. B. Gallagher. 1989. New initiatives for wildlife conservation: The need for movement corridors. In *Preserving communities and corridors*, ed. G. Mackintosh, 11-34. Washington, DC: Defenders of Wildlife.

Holling, C. S., ed. 1978. *Adaptive environmental assessment and management*. New York: John Wiley & Sons.

Horn, H. S. 1978. Optimal tactics of reproduction and life-history. In *Behavioral ecology: An evolutionary approach*, eds. J. R. Krebs and N. B. Davies, 411-429. Sunderland, MA: Sinauer Associates.

Hunter, M. L. 1990. *Wildlife, forests, and forestry*. Englewood Cliffs, NJ: Prentice-Hall.

Jackson, J. A. 1971. The evolution, taxonomy, distribution, past populations and current status of the red-cockaded woodpecker. In *The ecology and management of the red-cockaded woodpecker*, ed. R. L. Thompson, 4-29. Tallahassee, FL: Bureau of Fish and Wildlife and Tall Timbers Research Station.

Janzen, D. H. 1988. Management of habitat fragments in a tropical dry forest: Growth. *Annals of the Missouri Botanical Garden* 75:105-116.

Leopold, A. 1941. Wilderness as a land laboratory. *Living Wilderness* 6(July):3.

―――――. 1949. *A Sand County almanac*. New York: Oxford University Press.

Longino, H. 1990. *Science as social knowledge*. Princeton: Princeton University Press.

Lyon, L. J. 1983. Road density models describing habitat effectiveness for elk. *Journal of Forestry* 81:592-595.

Mader, H.-J. 1984. Animal habitat isolation by roads and agricultural fields. *Biological Conservation* 29:81-96.

Martin, R. E., J. B. Kauffman, and J. D. Landsburg. 1989. Use of prescribed fire to reduce wildfire potential. In *Proceedings of the symposium, Oct. 26-28, 1988, Sacramento, CA, Fire and watershed management*, N. H. Berg, technical coordinator, 17-22. Gen. Tech. Rep. PSW-109. San Francisco: USDA Forest Service.

Maser, C., R. G. Anderson, K. Cromack, J. T. Williams, and R. E. Martin. 1979. Dead and down woody material. In *Wildlife habitats in managed forests: The Blue Mountains of Oregon and Washington*, ed. J. W. Thomas, 78-95. USDA Agric. Handbk. 553. Washington, DC: USDA Forest Service.

Maser, C., R. F. Tarrant, J. M. Trappe, and J. F. Franklin, eds. 1988. From the forest to the sea: A story of fallen trees. Gen. Tech. Rep. PNW-GTR-229.

Portland, OR: USDA Forest Service, Pacific Northwest Forest and Range Experiment Station.

Mech, L. D., S. H. Fritts, G. L. Radde, and W. J. Paul. 1988. Wolf distribution and road density in Minnesota. *Wildlife Society Bulletin* 16:85-87.

Moldenke, A. R., and J. D. Lattin. 1990. Dispersal characteristics of old-growth soil arthropods: The potential for loss of diversity and biological function. *Northwest Environmental Journal* 6:408-409.

Morrison, P. H. 1990. *Ancient forests on the Olympic National Forest: Analysis from a historical and landscape perspective.* Washington, DC: The Wilderness Society.

Mutch, R. W. 1970. Wildland fires and ecosystems—a hypothesis. *Ecology* 51:1046-1051.

Myers, R. L. 1985. Fire and the dynamic relationship between Florida sandhill and sand pine scrub vegetation. *Bulletin of the Torrey Botanical Club* 112:241-252.

—————. 1990. Scrub and high pine. In *Ecosystems of Florida*, eds. R. L. Myers and J. J. Ewel, 150-193. Orlando, FL: University of Central Florida Press.

Norse, E. A. 1990. *Ancient forests of the Pacific Northwest.* Washington, DC: Island Press and The Wilderness Society.

Norse, E. A., K. L. Rosenbaum, D. S. Wilcove, B. A. Wilcox, W. H. Romme, D. W. Johnston, and M. L. Stout. 1986. *Conserving biological diversity in our national forests.* Washington, DC: The Wilderness Society.

Noss, R. F. 1983. A regional landscape approach to maintain diversity. *BioScience* 33:700-706.

—————. 1985. On characterizing presettlement vegetation: How and why. *Natural Areas Journal* 5(1):5-19.

—————. 1987. Protecting natural areas in fragmented landscapes. *Natural Areas Journal* 7:2-13.

—————. 1990. Indicators for monitoring biodiversity: A hierarchical approach. *Conservation Biology* 4:355-364.

—————. 1991. Wilderness recovery: Thinking big in restoration ecology. *Environmental Professional* 13:225-234.

—————. 1993. Wildlife corridors. In *Ecology of greenways,* eds. D. Smith and P. Hellmund. Minneapolis, MN: University of Minnesota Press.

Pickett, S. T. A., and J. N. Thompson. 1978. Patch dynamics and the design of nature reserves. *Biological Conservation* 13:27-37.

Platt, W. J., G. W. Evans, and S. L. Rathbun. 1988. The population dynamics of a long-lived conifer (*Pinus palustris*). *American Naturalist* 131:491-525.

Repenning, R. W., and R. F. Labisky. 1985. Effects of even-age timber management on bird communities of the longleaf pine forest in northern Florida. *Journal of Wildlife Management* 49:1088-1098.

Ripple, W. J., G. A. Bradshaw, and T. A. Spies. 1991. Measuring forest landscape patterns in the Cascade Range of Oregon, USA. *Biological Conservation* 57:73-88.

Schowalter, T. D. 1988. Forest pest management: A synopsis. *Northwest Environmental Journal* 4:313-318.

Swanson, F. J., J. F. Franklin, and J. R. Sedell. 1990. Landscape patterns, disturbance, and management in the Pacific Northwest, USA. In *Changing landscapes: An ecological perspective*, eds. I. S. Zonneveld and R. T. T. Forman, 191-213. New York: Springer-Verlag.

Thiel, R. P. 1985. Relationship between road densities and wolf habitat suitability in Wisconsin. *American Midland Naturalist* 113:404-407.

Thomas, J. W., R. G. Anderson, C. Maser, and E. L. Bull. 1979. Snags. In *Wildlife habitats in managed forests: The Blue Mountains of Oregon and Washington*, ed. J. W. Thomas, 60-77. USDA Agric. Handbk. 553. Washington, DC: USDA Forest Service.

Thomas, J. W., E. D. Forsman, J. B. Lint, E. C. Meslow, B. R. Noon, and J. Verner. 1990. *A conservation strategy for the northern spotted owl.* Portland, OR: USDA Forest Service, USDI Bureau of Land Management, USDI Fish and Wildlife Service, and USDI National Park Service.

Toumey, J. W. 1947. *Foundations of silviculture, upon an ecological basis.* 2d ed. New York: John Wiley & Sons.

Walker, J., and R. K. Peet. 1983. Composition and species diversity of pine-wiregrass savannas of the Green Swamp, North Carolina. *Vegetatio* 55:163-179.

Walters, C. J. 1986. *Adaptive management of renewable resources.* New York: McGraw-Hill.

Walters, C. J., and C. S. Holling. 1990. Large-scale management experiments and learning by doing. *Ecology* 71:2060-2068.

Whittaker, R. H. 1960. Vegetation of the Siskiyou Mountains, Oregon and California. *Ecological Monographs* 30:279-338.

Wiens, J. A. 1989. *The ecology of bird communities.* Vol. 2, *Processes and variations.* New York: Cambridge University Press.

Wilcove, D. S., C. H. McLellan, and A. P. Dobson. 1986. Habitat fragmentation in the temperate zone. In *Conservation biology: The science of scarcity and diversity*, ed. M. E. Soulé, 237-256. Sunderland, MA: Sinauer Associates.

Wisdom, M. J., L. R. Bright, C. G. Carey, W. W. Hines, R. J. Pederson, D. A. Smithey, J. W. Thomas, and G. W. Witmer. 1986. *A model to evaluate elk habitat in Western Oregon.* Portland, OR: USDA Forest Service, Pacific Northwest Region.

2

An Ecosystem Perspective
on Sustainable Forestry and New Directions
for the U.S. National Forest System

Hal Salwasser
New Perspectives, USDA Forest Service

Douglas W. MacCleery
Timber Management, USDA Forest Service

Thomas A. Snellgrove
Forest Harvesting Research, USDA Forest Service

Concerns about the sustainability of American forests have been growing for the past several decades. Specific issues include the ecological health of forests, their wildlife diversity, their resilience to stress and climate change, their productivity for wood and other resources, the environmental services provided by forests, and their aesthetics. Consequently, changes in forestry practices to address these concerns are occurring on all U.S. forest ownerships, most especially in the National Forest System. However, not all changes are the same. Some forests are being managed more intensively for wood and wood fiber, especially on private industrial lands (Bingham 1991), while others are being managed primarily for nonwood values and uses, principally on public lands. The changes in American forestry can be summed in one phrase: more diversity.

Forestry education and research are also changing but at a slower pace than changes in management direction (National Research Council 1990;

Lubchenko et al. 1991). Research and education are expanding to address new knowledge about the dynamics of forests as ecological systems and the connections between forest product technologies, forest management, economies at various geographic scales, and our society's changing values and needs.

The fundamental principles of forestry and natural resources management in the United States seem to be fairly stable: (1) science-based land stewardship, (2) efficiency in the production and conservation of natural resources, and (3) socially responsive management to meet landowner objectives. However, interpretations of these principles increasingly reflect broader concerns for the various uses and values of forests (Sample 1991).

While changes are under way in American forestry, public understanding of policy choices is not clear. It is often clouded by incomplete knowledge and distorted information. Media coverage of forestry issues, for example, tends to stress inflammatory portrayals of a "war in the woods" (Mitchell 1990), "rage over trees," the "end of nature" (McKibben 1989), or alleged mismanagement of public forests (Baden 1991; Knize 1991; Knudsen 1991). These portrayals reflect growing concerns over forests, but sensationalism rather than facts tends to dominate public debate.

Some Key Issues

Rhetoric aside, there are substantive issues in American forestry that merit attention, including the following.

- Value conflicts over what forest policies should favor. For example, should they favor certain wildlife species, old-growth forest ecosystems, or timber supplies in Pacific Northwest forests (Caulfield 1990; Thomas et al. 1990; Johnson et al. 1991)?
- Sustainability of desired forest conditions (Botkin 1990; Fri 1991; Gale and Cordray 1991; Sample 1991). For example, how can managers enhance biodiversity conservation, resilience to stress, productivity of desired goods and services, and aesthetics in forests that are now vulnerable to catastrophic wildfire due to historic fire suppression, prolonged drought, and insect epidemics?
- Clear-cutting to harvest trees and regenerate new forests (Greber and Johnson 1991). What harvest and regeneration technologies will be

more conducive to maintaining healthy forests and yet be economically feasible?

- Selective harvest effects in privately held forests in the East on forest diversity and productivity (Nyland 1991). Are loggers highgrading eastern forests again?
- Protection of endangered species that live in forest, such as spotted owls, large predators, red-cockaded woodpeckers, certain fish species, marbled murrelets, and some amphibians (Kohm 1991). Can endangered species be accommodated in forests that are managed for other values and uses as well?
- Fragmentation of forests into small and disjunct patches in places where forest cover was naturally contiguous and the hypothesized threats that such fragmentation poses to wildlife and fish species that have specialized habitat requirements (Terborgh 1990; Johnson et al. 1991). Is this fragmentation a serious threat to wildlife, or is it an artifact of the scales at which people analyze data?
- Economics of forestry decisions. For example, what can be done to reduce sales of timber that do not return revenues to cover costs on public lands (O'Toole 1988; Baden 1991)?
- Sustaining supplies of forest products (Sedjo and Lyon 1990) and their related job opportunities for people, especially in small, rural communities with little economic diversity.

Contrary to these concerns, some American forest types seem to have become too abundant or too mature for the long-term health of the land in large part because of the effectiveness of fire suppression. Examples are pinyon-juniper forests in the American Southwest, mixed conifer forests of the Sierra Nevada and intermountain West, and lodgepole pine in the Northern Rockies. Some reduction of forest cover, especially opening the canopy closure of trees to allow light and moisture to nurture an understory of plants that will better protect soils and watersheds, is the pressing management challenge. Simply letting fires burn unchecked to restore the natural roles of fire is rarely feasible or desired because of fuel situations and the risks of damaging soils under high heat or the proximity of human habitations. Also, some of the perceived forest fragmentation is actually forest patches returning in agricultural settings where little or no forest has existed for a century or more.

Meanwhile, as the real and perceived issues regarding forests are debated, people who depend on forests for their livelihood and identity are caught in a vicious cross fire of competing political agendas for who controls public and private resources and assets.

THE CENTRAL POLICY QUESTIONS

Ultimately, the wisdom of forest policy choices will depend on valid information regarding forest conditions and their capabilities as well as the economic and environmental ramifications of various management options. People need this information to answer two basic questions. These questions arc fundamental to all else in forest policy: What does a society desire for the current and future roles of forests in its communities, economies, and environments? Then, once that is decided, What are the most effective means to achieve and sustain these desired roles and conditions?

AN ECOSYSTEM PERSPECTIVE ON FORESTS

This chapter describes an ecosystem perspective on forest policy choices that has been evolving for several years in the USDA Forest Service. An ecosystem perspective is different from the historic American focus on producing and renewing selected resources (such as timber, game fish, and livestock forage) or single sectors of forest-related enterprises (such as wood products, recreation, and cattle industry). It is more holistic in adding a focus on relationships between various forest conditions, natural events and processes, various human uses of forests, and the different and changing values of forests to people at multiple geographic scales over time. It is further concerned with how natural events and processes together with human uses of forests in one place can effect environmental and economic changes over time and in different places. This reflects two fundamental ecological ideas, (1) that everything is connected to everything else in ecosystems, making it impossible to take only one action without causing a chain of other reactions (Hardin 1985), and (2) that all ecosystems are parts of larger ecosystems that receive their "externalities" and, in turn, set the context for conditions in the subsystems (Allen and Starr 1982).

An ecosystem perspective includes attention to selected resources and their related economic and social systems. Thus, it does not discard the knowledge and experiences gained through past accomplishments in forestry and other branches of resource management. It tries, however, to place them in a larger, more realistic context than the simple scientific notion that certain parts of ecosystems can be sustained or enhanced by decreasing diversity to focus on the productivity of those parts.

ABOUT THIS CHAPTER

We start the chapter with a general description of some basic relationships among people, forests, forest products, and environmental quality. We then describe several historical trends and influences that brought about current conditions of forests in the United States. This leads to a discussion of the capabilities of national forests and grasslands in the United States to partially address people's needs and concerns. Finally, we describe how certain policies and practices of the USDA Forest Service are changing to reflect the emerging ecosystem perspective on forestry and resource management. These new policies and practices grew out of a series of research and management projects carried out during 1990-92 under the title New Perspectives for Managing the National Forest System.

Do not expect comprehensive interpretations or perfect solutions to difficult choices from this chapter. These may not be possible for issues as broad and complex as those raised by policy choices for sustaining desired forest conditions, values, and uses. In any case, they are beyond the capabilities of these authors or of a single chapter or book. Our goals are merely to broaden the dimensions of ongoing discussions on what to do about our forests and to describe what the USDA Forest Service is doing to help find answers.

FORESTS, PEOPLE, AND THE ENVIRONMENT

Forests and forestry were prominent agenda items at the United Nations Conference on Environment and Development in 1992. Part of the reason for such high concern over forests is the role of forests in the global environment (Silver and DeFries 1990). Another is that forests are great sources of wealth and well-being for people (Marsh 1864; Toynbee 1976; Clawson 1979; Williams 1989; Perlin 1991). For example, forests are part of every generation's heritage of biological and cultural diversity. They are, thus, also part of the legacy that each generation will leave for the future. Forests are a natural factory for many renewable resources (e.g., wood, water, medicinal plants and animals, foods, and materials for shelter and clothing) (Frederick and Sedjo 1991). They are also vital organs of planetary health, holding soils, cleansing waters, and maintaining atmospheric balances (Silver and DeFries 1990). They are playgrounds. They are sources of livelihood and personal identity. And, for many people, forests are places for inspiration and communion with nature. In all regards forests are significant in the quality and material standard of human life. This is no more or no less true in the United States than it is in Amazonia, Europe, or India. Only the dimensions and magnitudes of relationships between people and forests and the challenges faced by forest stewards differ.

TRENDS IN FORESTS AVAILABLE FOR USE, ENJOYMENT, AND ENVIRONMENTAL SERVICES

Given that forests are so important, let us briefly review what is happening to them. Forests and woodlands now cover an estimated 31 percent of the planet's terrestrial surface (4.1 billion hectares; 10.1 billion acres) (World Resources Institute 1990). This is about 66 percent of the forested area that existed before the industrial revolution of the seventeenth and eighteenth centuries. Meanwhile, the number of humans has grown elevenfold since that time, from an estimated 500 million people to about 5.5 billion.

In per capita terms, this means that there were about 12 hectares (30 acres) of forest per human being on the planet in 1750. In 1990, however, there was only about 0.75 hectare (1.8 acres) per person on average.

Forests in the United States now cover about 32 percent of the nation's area (296 million hectares; 731 million acres) (Haynes 1990). This is also about 66 percent of the forest cover that existed before European settlement (Clawson 1979). Approximately 150 million hectares (370 million acres) of the original forest have been converted to other uses, mostly agriculture, while some previously converted lands have returned to forest cover during the twentieth century.

Meanwhile, the human population of what is now the United States has grown by 25 times since the 1600s: from an estimated 10 million before introduced diseases decimated Native Americans to 250 million in the latter part of the twentieth century. In per capita terms, this equates to a change from 45 hectares (110 acres) of forested area per human inhabitant in the 1700s to an average of 1.2 hectares (2.9 acres) per person in 1990.

Declines in the ratio of forested area per human during the past four centuries tell us several things. One is that people must have become far more efficient and conservative in obtaining and sustaining forest products and forests in general or there probably would not be any forest left. Absent some form of protection and management, world and national forests would probably have been reduced much further given the increase in human population. Obviously, from these data, there is now less potential forest area available for each current and future human to occupy as a residence, in which to find or produce resources, or to provide for various environmental services. Of course, this situation applies to many land, water, and biotic resources of the planet (Silver and DeFries 1990). Yet, there is a paradox in this trend. The growing human population, with its intellectual capacity and ingenuity, is and will continue to be both a pressure on the environment and the source for improving environmental quality and the standard of living for people.

DIFFERENCES IN HOW HUMANS USE FORESTS

Resolving this paradox is not simple. Solutions in one place or time may not work well in others because relationships between people and forests are not static. They differ from place to place, and they change dramatically over time. As recently as five centuries ago, people in what is now the United States lived in tight association with forests. Today, the United States is an urban society. More than 90 percent of the population lives in cities or suburbs. Many lack direct contact with the land and have for more than a generation. They may visit the land as a recreating tourist, but few experience the land as a resilient, renewable, and productive resource as did many of their parents and grandparents. Only those who hunt, fish, or gather berries, fuelwood, or other materials for their use experience in any direct way their ecological roles in forest systems. Many urban people, including those with relatively high intelligence and education, do not know or choose to ignore the fact that the land is where their food, shelter, and water ultimately come from. Perhaps daily living in congested, highly modified surroundings shapes such a view of the world. Whatever the cause, urban people typically have little personal experience upon which to judge the veracity of information they receive on forests from newspapers, mail-order campaigns of various political advocacy groups, magazines, or television.

People in rural areas or developing countries still know their relationship with forests firsthand. They use forests to meet the basic needs of existence (Marsh 1864; Thomas 1956; Toynbee 1976; Perlin 1991), just as many people did in the United States until well into the twentieth century (Clawson 1979; Williams 1989). In some cases, this subsistence relationship has been maintained for centuries, perhaps even millennia. In other cases, however, food, shelter, medicines, and fuels were taken until the forest and its soils disappeared or until economies developed to the point where people could find their basic resources elsewhere and, thus, afford to conserve forests for other values.

In some countries, such as India, about 70 percent of the wood harvested from forests is used for domestic energy (Lal 1989). When a human population that depends daily on forest resources becomes too dense, such uses can have devastating effects on forest sustainability.

Wood was once the primary energy source in the United States also, but this has changed over time. In 1850, wood provided about 95 percent of U.S. domestic and industrial energy measured as British thermal units (BTUs) (Fedkiw 1989). Today, the United States uses fossil wood and fossil animals in the form of oil, coal, and natural gas for most of its energy.

Forty-four percent of all wood used in the United States from 1980 to 1988 was for construction (lumber, plywood, and veneer); about 27 percent was for pulp and paper; and 22 percent was for fuelwood (data from Table 4 of Ulrich 1990). From 1980 to 1988, the total production and consumption of wood increased by an average of 0.3 and 0.4 percent per year, respectively. Interestingly, after declining in midcentury, the amount of wood used for fuel in the United States during the 1980s approached the amount used in the early 1900s.

Changes in how wood is used in the United States and their implications for forest conditions are directly tied to changes in our economic prosperity, lifestyles, and technologies, yet the linkage between material standard of living in the United States and management of natural resources remains poorly understood by many people. This is especially true concerning the role of forest management in providing both environmental and economic benefits.

WOOD CONSUMPTION AND SUPPLY AS FOREST INFLUENCES

The use of wood is a major factor in forest sustainability, perhaps the dominant one (Clawson 1979; Williams 1989; Perlin 1991). It will probably remain so for a long time to come. Both global and U.S. wood

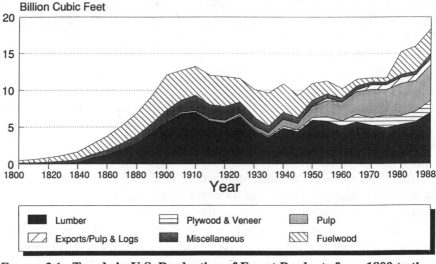

FIGURE 2.1. **Trends in U.S. Production of Forest Products from 1800 to the 1980s.**
Data from Sedjo (1990) and Ulrich (1989, 1990).

production and use continue to rise (figure 2.1). The United States produces about 25 percent and uses about 33 percent of world industrial
roundwood production (all wood used other than for fuel). We use about
50 percent of world paper production (Haynes and Brooks 1991). The
United States is the biggest total user of wood and the biggest per capita
user (Postel and Ryan 1991).

United States per capita use of wood other than for fuel is on the order of
1.5 times that of other industrial nations and as much or more than 100 times
that of some nonindustrial nations (Postel and Ryan 1991). And it is not
declining. United States wood production and consumption increased by
about 28 percent from the 1970s to the 1980s, primarily because of increased use of wood in construction, home heating, and wood-fueled processes in the forest products industry (table 2.1; Ulrich 1990).

Between 1980 and 1988, total net timber product imports were 9 percent
of total U.S. consumption, making the United States a net importer of
wood products by an average of 448,000 cubic meters (1.58 million cubic
feet) per year. Net imports averaged 12 percent of total use for wood pulp
and 16 percent for lumber and plywood products during this period
(Ulrich 1990). About 27 percent of the lumber used in the United States
during the late 1980s was imported from Canada.

TABLE 2.1. **Total U.S. Consumption of All Timber Products from 1950 to 1988, Including Fuelwood**

| Years | Million Cubic Meters Roundwood Equivalent | |
	Total	Per Capita
1950–54	340	2.2
1955–59	337	2.0
1960–64	332	1.8
1965–69	362	1.8
1970–74	372	1.8
1975–79	411	1.9
1980–84	466	2.0
1985–88	542	2.2

Data from Ulrich (1989), Table 4; Ulrich (1990), Table 4; and U.S. Departments of Agriculture and Commerce.

The United States recently produced about 23 percent of the softwood sawtimber it used annually from trees harvested in national forests. This percentage is declining due to policy choices to protect public forests for other values, such as watersheds, aesthetics, and wildlife.

It is not clear whether Americans will use less wood as production from public forests declines or whether they will find their wood elsewhere, such as from U.S. private lands, as assumed by the recent Resources Planning Act Program for 1990 (USDA Forest Service 1990), or the forests of other nations. If the increasing trend in total wood consumption is not reduced or offset by increased production from domestic sources, U.S. wood consumption could put increasing pressures on the forests of other nations.

MATERIAL ALTERNATIVES TO WOOD

Using less wood for all of the things wood is used for is one policy option to reduce the pressure on forests. Many nonwood products are used in concert with wood or have the capability to substitute for wood in construction applications. But these substitutions are not benign; they have both environmental and economic consequences (table 2.2) (Koch 1991; Alexander and Greber 1991; Bowyer 1991a, 1991b, 1991c).

TABLE 2.2. **Estimated Energy Required in the Manufacture of Various Wall Systems for Building Construction**

Type of Wall	Energy to Construct 100-square-meter Wall[a]	
	Million BTU Oil Equiv.	Equiv.
Plywood siding, no sheathing, 2×4 frame	21	1.0
MDF siding, plywood sheathing, 2×4 frame	27	1.3
Aluminum siding, plywood, insulation board, 2×4 frame	53	2.5
MDF siding, plywood sheathing, steel studs	55	2.6
Concrete building block, no insulation	184	8.8
Brick veneer over sheathing	193	9.2

Basic data are from CORRIM (1976) as cited in Bowyer (1991c) and converted to metric equivalent.
[a] Estimates include energy consumption involved in logging (or extraction), manufacture, transport to the building site, and construction.

Compared to its alternatives, wood is an environmentally superior material. It is virtually the only *renewable* resource that is widely and economically suitable for structural and architectural purposes (Koch 1991). Alternatives to wood in construction uses—steel, aluminum and other metals, concrete, and plastics—are not renewable (although they are recyclable at varying energy costs). They use considerably more energy per unit of production than does wood. Other natural alternatives such as adobe and rock are not universally available.

Koch (1991) estimated that steel studs require about 9 times more energy to produce and transport to a construction site than do wood studs; aluminum siding requires 4 times more energy, and brick veneer 22 times more energy compared to equivalent wood siding. Concrete floors use 21 times more energy to produce than do wood floors. And wood is generally more energy efficient in use than many manufactured substitutes, paying energy dividends throughout the life of a building.

Because the amount of wood used annually in the United States is approximately equal by weight to the combined amount of all potential wood substitutes (Bowyer 1991a), replacing a significant percentage of wood with manufactured substitutes could have substantial effects on national energy consumption and global carbon dioxide emissions. Koch (1991) estimated that, for each 5.7 million cubic meters (1 billion board feet) of wood replaced with manufactured substitutes, annual energy consumption could increase by about 2.7 billion liters (720 million gallons) of oil and carbon dioxide emissions could increase by 6.8 billion kilograms (7.5 million tons). This amount of wood is about 1.3 percent of U.S. annual production.

The actual energy efficiencies of wood and alternative materials probably vary according to manufacturing technologies and patterns of material use. Many of the possible substitutions may not occur for various reasons. The essential points are that (1) wood is a significant material in the United States economy and (2) there are energy costs and environmental implications associated with substituting other materials for wood if such substitutions are a response to policy choices on forest sustainability.

A HISTORIC PERSPECTIVE ON FORESTS IN THE UNITED STATES

In the current debate over American forests and the environment, it would be easy to think that this is the first generation to care about future wood supplies, wildlife diversity, or forest sustainability. A look at recent history should dispel this belief (Fedkiw 1989).

In the latter half of the nineteenth century, the nation's human population was expanding rapidly. Settlement of the American frontier was seen as a laudable national objective. One unfortunate consequence of this settlement was depletion of much of the nation's forests and wildlife. To feed a rapidly growing population, Americans cleared forests for farmland on a massive scale. Increasing urbanization and industrialization also created a huge demand for lumber and structural timbers to build growing cities. Large-scale logging, especially in states around the Great Lakes of the Midwest, was followed by sometimes massive wildfires (Williams 1989). Nationally, forest growth rates were a fraction of harvest levels.

Wildlife was also under assault during the late 1800s (Trefethen 1975; Dunlap 1989). Firearms technology had improved dramatically. Game laws were nonexistent or poorly enforced. There was virtually unrestricted market hunting of all kinds of wildlife for food, furs, and feathers (which in the late 1800s were in great demand for women's hats), as well as habitat modification caused by forest clearing for farms, logging, and wildfire.

By 1900, populations of many wildlife species were so depleted that they would have been on an endangered species list had one existed. These include now common game animals, such as white-tailed deer (which had been extirpated entirely from most eastern states), wild turkey, pronghorn antelope, moose, bighorn sheep, and many smaller mammals whose pelts were valuable for furs. Many waterfowl, including swans and wood ducks, Canada geese, and plumed wading birds such as herons, egrets, and ibises, were also on the brink of extinction. Many other wildlife species, although not in danger of extinction, were much depleted.

These conditions were not created intentionally. They were consequences of actions taken for other reasons. People generally do what they think is necessary to feed themselves and their families and to build their communities and nations. What happened to forests and wildlife during the late 1800s and into the early 1900s in the United States was an extension of a long settlement history that was accelerated by European immigrants in the early 1600s. By the 1870s, the effects of settlement were magnified by a rapidly expanding population and advancing technology.

What is important from today's perspective is that there were farsighted individuals in the latter half of the nineteenth century, including political leaders such as President Theodore Roosevelt, who recognized that new approaches were needed. They took bold actions to put these new approaches in place (Trefethen 1975).

The new ideas were called conservation, and they set the stage for the emergence of the nation's first environmental movement. The policy framework for conservation that emerged during the early 1900s emphasized protection of forests from wildfire and of wildlife from overharvesting. Specific actions focused on (1) the acquisition of scientific knowledge through research on forest and wildlife culture and management and its enlightened application by resource professionals, both public and private; (2) the promotion and encouragement of the protection of forests, regardless of ownership, from wildfire, insects, and disease; (3) the encouragement of the productive management of private forest lands through tax incentives and technical and financial assistance; (4) the adoption and enforcement of strong state and federal wildlife conservation laws; and (5) the acquisition and management of public lands for both commodity and amenity uses and values. A key element of the public policy framework

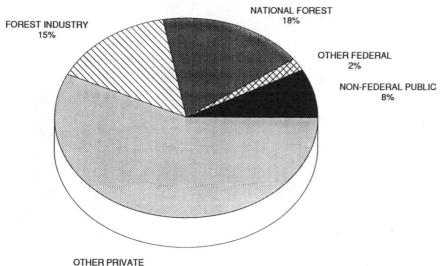

FIGURE 2.2. **Timberland Ownership as a Percentage of the United States Total in 1987.**
Data from Haynes (1990).

was strong cooperation among federal, state, and private sector interests to achieve common goals (Fedkiw 1989).

Together with technological improvements in agriculture that allowed millions of hectares of farmland to revert to forests (Fedkiw 1989), these policies resulted in a general and dramatic recovery of American forests and wildlife (MacCleery 1991; Shands 1991).

FOREST CONDITIONS AND WOOD USE TODAY

Today, U.S. forests and forest wildlife have recovered substantially from low points at the turn of the century. In aggregate, forests in the United States have increased in productivity throughout the latter part of this century, a direct response to forest and conservation policy choices and technology improvements (MacCleery 1991).

About 20 percent of the 196 million hectares (483 million acres) of total U.S. timberland area—these are forests capable of producing 1.4 cubic meters of industrial roundwood per hectare per year (20 cubic feet per acre per year) and not reserved for uses that preclude timber harvest—is now managed by agencies of the federal government; 15

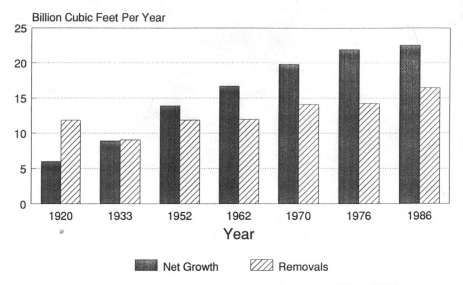

FIGURE 2.3. **U.S. Timber Growth and Removals from 1920 to 1986.**
Data from USDA Forest Service (1973, 1982) and Haynes (1990).

percent by forest industries; 8 percent by agencies of state and local governments; and 57 percent by nonindustrial private landowners (figure 2.2). Nationwide, wood growth on these lands, which in 1900 was a fraction of wood removal, attained a general balance with harvest by the 1940s. In recent decades average timber growth has continually exceeded harvest (figure 2.3).

The total volume of wood in U.S. forests is now 25 percent greater than it was in 1952. Actively growing forests are great carbon "sinks." American forests are estimated to sequester the equivalent of about 9 percent of total carbon dioxide emissions from all sources in the United States (M. Fosberg, personal communication to T. A. Snellgrove, 1992).

Tree planting was at record levels throughout the 1980s. More than 10.5 million hectares (26 million acres) of trees were planted during the 1980s. This is an area of land the size of the state of Virginia. Last year for every child born in the United States, more than four hundred trees were planted.

The area of forest burned by uncontrolled wildfire has been reduced from a range of 16 to 20 million hectares (40 to 50 million acres) annually in the early 1930s to a range of 0.8 to 1.6 million hectares (2 to 4 million acres) today. We need to rethink the roles of fire in keeping wildland ecosystems diverse and resilient to stress. But investments in reforestation, forest management, and forest fire control have resulted in more forested area with tree growth in the United States today, about three and a half times what it was in 1920 (Fedkiw 1989). This is a significant accomplishment because the first step in any forest conservation strategy is to keep forest areas in forest cover.

Because American forests are so abundant and productive, the nation can and is protecting more forests, both native and restored, for environmental services, aesthetic values, and amenity uses. For example, the area of forest land in parks, wilderness areas, and similar reserves in which timber harvest is prohibited has increased significantly in recent years. As of the late 1980s, about 14 million hectares (34.5 million acres) of forest that is biologically capable of sustaining timber production have been designated for nontimber values and uses in such areas as parks and wilderness in the United States (Haynes 1990), nearly double the area in such designations in 1970 (MacCleery 1991). This is an area the size of the state of Florida.

Harvest and manufacturing efficiencies have also shown significant improvements since the turn of the century. Early records are sketchy, but since the 1950s logging residues have decreased by an estimated 10 percent for softwoods and about 40 percent for hardwoods. Although consistent statistical data are not available, it is also well accepted that utilization of trees killed by fire, insects, or diseases has also increased dramatically. Further, the proportion of harvested trees that is effectively converted into lumber or veneer has increased by about 20 percent in sawmills and about 22 percent in plywood plants (Haynes 1990). Advanced technologies such as thinner saw blades, electronic measurement systems, and computer-assisted milling have all contributed to improved fiber recovery.

New technologies for increasing the usable wood from a forest (improved utilization), for extending the life of wood in use (preservation), and for using the same fiber several times (recycling) have taken significant pressure off U.S. forests for raw material. All of these efforts have reduced by hundreds of thousands of hectares the area of annual harvest that otherwise would have supplied the United States with wood products.

There are opportunities for even more improvement in wood utilization and recycling (Ince and Alig 1991), and this must be an integral part of new policies for improving forest sustainability. Postel and Ryan (1991) estimated a potential for conservation technologies to reduce raw material demand by up to 50 percent. Their work may identify what is technically possible if people decide to take action to increase the conservation of wood resources. Although it seems that pressure to recycle paper is coming as much from a local shortage of landfill space as it is from a new sense of global citizenship, it is coming nevertheless. Using conservation technologies to reduce per capita use of raw material could help take pressure off forests until population growth again raises total demand.

FOREST WILDLIFE TODAY

Several species of American wildlife, such as the passenger pigeon and Carolina parakeet, became extinct in response to forest changes and human uses during this century. Many others that were poised on the brink of extinction in 1900, however, staged remarkable comebacks. Because of actions that were set in motion in the early decades of this century, many forest wildlife species are both more abundant and more widespread today than they were in 1900 (Thomas 1989).

The pattern that emerged after the 1930s was a substantial recovery in the numbers and range of forest wildlife that can tolerate a relatively broad range of habitat conditions, so-called "habitat generalists." Fortunately, many U.S. forest wildlife species are habitat generalists. One reason may be the natural dynamics of North American forests and the frequency of disturbance in the natural regime (Williams 1989; Botkin 1990).

Saying that many wildlife species have staged remarkable comebacks, however, does not imply the absence of problems. Species with specialized habitat requirements are increasingly of concern today, including the following examples.

- The red-cockaded woodpecker and gopher tortoise, which are natives of fire-created southern pine savannas and woodlands
- The Kirtland's warbler, which is native to young jackpine forests in Michigan
- The spotted owl, which occupies mature and old-growth forests in the West

Many forest wildlife species, such as grizzly bears, wolves, elk, and some forest-interior birds, need large, contiguous areas of habitat. Some require very old and ecologically diverse forests. But not all habitat specialists are threatened by loss of old-growth or "ancient" forests. Some require active management of young forests for their survival, for example, Kirtland's warbler (Botkin 1990). Others, although needing mature forests, require specific habitat conditions, such as open savannas and woodlands, that are created by frequent ground fires, for example, the red-cockaded woodpecker. Even the old-growth Douglas fir forests required by the northern spotted owl are subclimax forest types that may eventually move toward different forest conditions without occasional, stand-replacing wildfires. Providing for the needs of habitat specialists will require purposeful and often active forest management to recreate or maintain desired conditions and processes, although not always for early-successional habitats.

FRAMING FOREST ISSUES AND POLICY CHOICES

Given the foregoing, let us now explore some factors that must be addressed to sustain forest conditions that will provide for desired uses, values, and environmental services.

THREE CHALLENGES

The first challenge in sustaining desired forest conditions, uses, and values is to understand better the dynamic relationships among people, forest resources, environmental services provided by forests, and overall standards and quality of human life. We only briefly touched on these from a historical perspective in this chapter. The subject of human-forest relationships is ripe for further research and education. The second challenge is to articulate and frame our understanding of human-forest relationships so that people can make informed choices through their democratic decision-making processes. Part of this challenge is the development and presentation of accurate and timely information on conditions of forest ecosystems and the estimated consequences of policy choices. The third challenge is to develop, demonstrate, and apply technologies for protecting, restoring, and enhancing the conditions, uses, and values of forests desired by communities of people from local to global scales.

FACING THE POPULATION DILEMMA

Demographers tell us that there will soon be six billion people living on Earth and perhaps as many as eight to ten billion by the middle of the twenty-first century. In the United States, the human population will surpass 300 million in several decades.

Therefore, reducing per capita consumption of natural resources by increased conservation and recycling, although critical in the short run to both economic and environmental goals, must be augmented by policies to curb population growth and improve the productivity of resources needed to support human existence. More people means that more resources will be consumed. Whatever the levels of future resource consumption, those resources will come from somewhere (Chappelle and Webster 1991). Corresponding with more consumption, more pollution may be produced and it must be dealt with in some way.

On the positive side of this challenge, the United States today has almost four times the human population it had a century ago, living at a substantially higher material standard of living. Yet our forests and wildlife are, in many ways, in significantly better condition today than they were in 1890. This provides reason for optimism. Contrary to the simpleminded equation of environmental impact being equal to population times affluence times technology (Ehrlich 1990), there is abundant evidence that affluence and technology are also factors that can offset the potential environmental effects of human population growth. The current abundance and diversity of U.S. forests that have greatly expanded the range of choices available for forest and wildlife conservation are conditions related to the affluence and technological capacity of Americans and to the conservation policy choices they made in the past.

Forest policy choices made by this generation will also influence future economies and environments, probably in ways that we do not envision. But people are ill-served when coverage of environmental issues is framed so narrowly as to make it impossible for them to understand the full dimensions of the choices available or even what those choices are.

OLD-GROWTH FORESTS AS A POLICY EXAMPLE

A case in point is the conservation of remaining stands of old-growth forest in the Pacific Northwest. Americans have been told by environmental advocacy groups through direct mail, newspapers, television, and magazines that the last remnant stands of old-growth forests on public lands are about to be logged and will be gone within a decade or two, precipitating an environmental disaster of unprecedented proportions. Since this is not true, someone is obviously lying to the public to advance a particular agenda. Wise policy choices cannot be made upon false and distorted information.

It has become increasingly clear in recent decades that old-growth forests have many unique environmental, economic, and spiritual values. It is also well known among conservationists that U.S. policies for protecting and sustaining old-growth forests have evolved from their origins in forest restoration in the East and the designation of national parks and wilderness areas in national forests throughout the nation several decades ago. Current policies now include specific provisions to restore and maintain old-growth forests as part of the overall forest landscape in every national forest and on other public lands as well.

National forests alone contain about 12 to 14 million hectares (30 to 35 million acres) of old-growth forests, an area about the size of the state of Florida. More than half of this old-growth forest is protected in wilderness areas and other land-use designations that do not permit timber harvest.

In the Pacific Northwest—Oregon, Washington, and Northern California—where old growth has been a particular issue, about 2.6 million hectares (6.3 million acres) of old growth remain on national forest lands—an area of old growth larger than Massachusetts and Rhode Island combined. It has been estimated that this represents 10 to 15 percent of the original mature and old-growth forest of the region. More than half of this old growth is also protected in wilderness and other land uses that do not permit timber harvest.

At timber harvest rates projected in current national forest plans, about 2.3 million hectares (5.6 million acres) of Pacific Northwest old growth will remain in ten years. After these forest plans are revised to include additional protection for the northern spotted owl (Thomas et al. 1990) and other forest values (Johnson et al. 1991), it is likely that substantially more than 2.3 million hectares of old growth will remain after ten years. These are the basic inventory facts.

Policy choices are implied but not obvious from these facts. The United States can harvest more trees from old-growth forests. It can save all that remain. It can even accelerate the development of mature forests into an old-growth condition. These are all choices that have been made in national forest plans. But choices on what to do about remaining old-growth forests must combine local and regional biological facts with political, social, and economic dimensions and consider at least national if not global implications of policy options.

Because the Pacific Northwest is a major player in national and global timber supplies (Sedjo and Lyon 1990), there are also national and perhaps global implications to choices concerning old-growth forests in the Pacific Northwest. The wood these forests might have produced will either come from somewhere else or be replaced by nonwood materials if the United States does not reduce its total wood consumption accordingly. This is not just because the U.S. timber industry wants to make a profit. It is because a lot of American people use a lot of wood. If they did not use wood there would be no timber industry. As you recall from previously presented data, Americans show little sign of changing this wood-use behavior.

Thus, it is important to consider how choices for protecting and managing old growth in the western United States might affect local and regional economies and environments there and elsewhere, including possible effects on national and global timber supplies, energy and greenhouse gas implications of using substitutes for wood, and effects on biodiversity from increased timber harvests in other regions and nations. We may reach the same decisions for sustaining old-growth forests in the Pacific Northwest as if we had not considered these factors, but at least we will be more aware of the larger dimensions of our choices.

The old-growth case cited here is just one example of using an ecosystem perspective in thinking about policy choices. There has been little success in obtaining public understanding of the full dimensions of forest policy choices in the United States. This difficulty is significant and probably due in large part to the inherent complexity of forest values. It is also due in part to the propaganda and political positioning for influence in a democratic society of various interest groups.

Further complicating policy choices is the fact that it is not always necessary to select either/or, protection or production options that might satisfy the particular agendas of politically strong groups. For example, forest management options to supply combinations of environmental and economic goals in many circumstances exist and are being refined (Gillis 1990).

COMPLEXITY AND SCALE

Forest policy choices are complex. They integrate many social values and needs as well as biological knowledge. Before answers can be found, questions must be posed correctly (Clark and Stankey 1991). The correct questions are rarely of a single dimension. Thus, policies to sustain ecologically diverse and productive forests will not often, if ever, be found in single-dimension choices, such as preserve public forests and produce more wood from private lands, or save this or that species, or promote this or that industrial development. As the ecologist Garrett Hardin (1985) said, it is not possible to do only one thing in an ecosystem because of the degree of interconnections, most of which are unknown.

This brings us to the issue of scale, perhaps the major issue in an ecosystem perspective on forest sustainability. Temporal and spatial scales are critical factors in framing questions and choices on forest policies. National and global perspectives are needed in addition to local and regional perspectives because in the long run it may do little good to conserve biodiversity in local and regional ecosystems if resultant increased human consumption of resources depletes the same somewhere else. Local action is always necessary. But global responsibility requires that local actions also be positive on national and global scales. The protection of backyard environments cannot ignore global ecological dynamics (Botkin 1990; Bowyer 1991a, 1991c; Brown 1992). *Act locally but think globally* is not merely a cliché. And it is harder to do than many people think.

What people do to protect forests or produce forest resources in their backyards affects their economic well-being, environmental quality, and biological diversity. What people do not do in their backyard forests also affects these things. But it does not stop there. What people do or do not do in their backyards also affects someone else's economic prosperity and environmental quality. This is because markets and environments are global.

There are global ramifications to U.S. consumption of fossil fuels, worldwide use of chlorofluorocarbons, the Green Revolution of the 1960s, public health advances on human morality factors, wilderness area designation, and high-yield silviculture. Whether these are positive or negative influences on forests, biological diversity, or the quality of human life depends on the temporal or spatial scale of analysis. That is, it depends on how broadly and how long we view our goals for sustaining desired conditions, uses, and values of ecological systems, including forested ones.

In addition to looking at the inner workings of forests to determine how to sustain them as ecological systems, which is essential but not sufficient, we must also look outside the forest to understand the inner workings of the larger ecosystems that set the context for forests: the regional, national, and global human societies (Clark and Stankey 1991), economies (Binkley 1991), and environments that forests influence (Botkin 1990; Silver and DeFries 1990).

LINKING PEOPLE, FORESTS, WOOD, WILDLIFE, AND CONSERVATION

Given a large and growing human population, its dependency on wood as an environmentally superior, renewable natural resource, a large global wood supply (Sedjo and Lyon 1990), a large capacity to grow more wood in managed stands (Bingham 1991), and the potential for conservation to reduce demand for raw materials (Postel and Ryan 1991), people may reasonably question continued cutting of native old-growth forests, including those of other nations. But growing and cutting trees in ways that leave soils, waters, and ecosystems in diverse and productive conditions is essential for both environmental and economic well-being. This proposition was affirmed by the Global Forestry Principles of the United Nations Conference on Environment and Development.

Thus, forest policy choices must occur within the intersection of goals for human communities, economic development, and environmental quality (figure 2.4). They must take into consideration interregional, international, and intergenerational transfers of benefits and costs in addition to immediate and local benefits and effects. For example, at some geographic scale, perhaps global but certainly national, and at some temporal scale, perhaps future generations but certainly future decades, the ethics and responsibility of consuming amounts of natural resources that a community of people is capable of producing with minimal undesired consequences but chooses not to should be called into question if, by failing to meet their needs domestically, they export the effects of their use to other places or future times or add more burden to the global environment through the use of less environmentally benign materials (Bowyer 1991a). This is a long-winded way of saying that environmental or social ethics cannot be defined solely on the basis of one's behaviors in one's immediate environment.

This issue of spatial and temporal transfers of the costs and benefits of forest policy choices highlights one of the major challenges in crafting

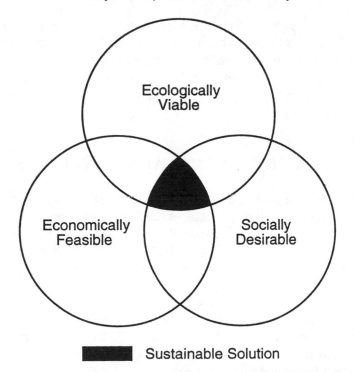

Sustainable Solution

FIGURE 2.4. **Ecosystem Management.** *Sustaining desired ecosystem conditions requires that management goals and actions fall within the intersection of three spheres: that they be simultaneously ecologically viable (environmentally sound), economically feasible (affordable), and socially desirable (politically acceptable). If the balance among these three criteria is not reasonable, there is a high likelihood that desired conditions will not be sustainable because of failures in one or more of the spheres.* After Zonneveld (1990).

policies and practices for sustainable forestry. Sustainable forestry is more than saving old growth, timber-industry jobs, spotted owls, roadless areas, endangered species, or even biodiversity. These are important factors in eventual solutions to what sustainable forestry is, but they may also be diversions from the real challenge because they focus attention on a still too narrow set of goals. What is fundamentally different between deciding to manage forests to sustain maximum output of timber and managing to sustain maximum output of spotted owls, for example, other than the fact that one addresses economic goals primarily and the other environmental goals? Neither addresses the intersection of both these legitimate goals. The ecosystem challenge is this: to sustain desired forest conditions, uses, and values in the context of global markets and environments and with sensitivity to local and regional equity in meeting human economic and social needs.

THE CAPABILITIES OF THE NATIONAL FOREST SYSTEM

Let us now turn to the largest single forest ownership in the United States as one example of how policy choices for forest ecosystems are evolving. The national forests and grasslands of the United States are owned by the people and managed in trust by the USDA Forest Service under direction from the U.S. Congress. They encompass 77 million hectares (191 million acres), about 8.5 percent of the land area of the United States, an area the size of the states of Texas and Louisiana combined. They are managed under a multiple-purpose mandate for both natural resource products, such as timber, livestock grazing, fishing, and minerals, and amenity uses and values, such as wildlife, recreation, nature study, and wilderness. They also serve a watershed protection role.

Because of the broad, multiple-use management mandate, there is considerable pulling and tugging in the forum of public debate over current and future roles of national forests and grasslands. Their current roles, both in the economy and as a source of natural and amenity values, are significant. For example, U.S. national forests and grasslands provide

- more than 70 percent of the nation's Wild and Scenic River System and 84 percent of its Wilderness Preservation System in the lower 48 states;
- watersheds that encompass one-half of the West's water supply, 5 percent of the East's water supply, and one-half of the nation's cold water fishery;
- more than 40 percent of all federal outdoor recreation, twice that of the national parks;
- habitat that supports about 70 percent of the vertebrate species richness that occurs in the nation, including most of the nation's ungulate species and more than two hundred threatened or endangered plant and animal species;

- about one-fifth of the nation's annual supply of softwood timber, which in the late 1980s generated $1.3 billion in annual receipts, 100,000 timber-related jobs, and more than $3 billion in timber-related income, primarily in local communities; and
- minerals that generated $4.9 billion in private sector revenues in 1990 and much of the current unexplored potential for oil, gas, and minerals in the United States.

RECENT TRENDS IN NATIONAL FOREST TIMBER HARVESTING

Timber production in the national forests is a major public issue, perhaps the most significant issue. About 70 percent of the National Forest System is covered by forest vegetation (54 million hectares; 133 million acres). About 30 percent of this forested area is classified as suitable for timber production (23 million hectares; 57 million acres). In these areas timber harvest is permitted as one of the multiple-use objectives, along with wildlife, recreation, grazing, watershed protection, and other uses and values. About 2.9 million hectares (7 million acres) of the land available for timber production is of the highest productive quality for growing wood (site I and above).

There are 31 million hectares (77 million acres) of forest land not available for timber production in the National Forest System, 58 percent of the total forest area. These forests continue to change annually under natural processes such as vegetative succession, most of which lead to a maturing forest, and disturbances such as wildfires, storms, droughts, and disease epidemics, most of which induce earlier successional stages of forest.

During the years 1984 to 1991, timber harvest for the purpose of producing forest products and regenerating a new forest (clear-cutting, removal cuts, and selection harvests) occurred on an average of 138,000 hectares per year (342,000 acres). This is about 0.6 percent of the area available for timber production in the National Forest System. Reforestation through planting and natural methods occurred on an average of 183,000 hectares per year (453,000 acres) during this period. Intermediate harvests for improving timber stand conditions or salvaging dead, dying, or diseased trees occurred on an average of 130,000 hectares per year (320,000 acres). This is slightly less than 0.6 percent of the area available for timber production. Intermediate harvests do not require reforestation because a growing forest is in place after harvest. Timber harvest for special purposes such as scenic vistas and campground safety and the preparation for future seed tree or shelterwood harvests accounted for an average 9,000 hectares of harvest per year (23,000 acres).

Nationally, the annual growth of wood on national forest timberlands continually exceeds wood harvest. For example, growth exceeded removal by about 55 percent in 1986. Because wildfires are controlled and the growing stock of timber is kept in relatively productive condition on National Forest System timberlands, net annual growth of timber stock (the annual growth of timber volume minus the losses through mortality and cull volume) has increased by about 67 percent since 1952, from about 65 million cubic meters to 97 million cubic meters in 1986 (Haynes 1990). In roundwood equivalents this is an increase in net growth from 9 billion to more than 13 billion board feet per year.

Annual timber harvest volume from the national forests has declined by about 24 percent in recent years, from about 86 million cubic meters per year from 1984 to 1989 (11.7 billion board feet roundwood equivalent) to 65 million cubic meters in 1991 (8.9 billion board feet roundwood equivalent). Forest plans identify a sustainable yield on the order of 75 million cubic meters per year (10.5 billion board feet roundwood equivalent). But political and scientific uncertainties make future national forest timber harvest levels unstable and unpredictable.

POLICY CHOICES FOR NATIONAL FORESTS

In recent years the growing urbanization, affluence, and mobility of Americans have caused a virtual revolution in expectations and demands placed on U.S. forests (particularly public forests). Demand for just about every conceivable use or value of the National Forest System continues to rise (USDA Forest Service 1990). Some of the trends pose direct conflicts between new expectations and traditional forest values and uses.

Resource assessments show that national forests and grasslands can produce more wood products to meet domestic and foreign needs. They can provide more recreation to enhance local economies and leisure time. They can focus on the recovery and conservation of endangered wildlife species. They can emphasize the conservation of biological diversity. They can provide more range forage for livestock. They can provide more access to minerals and energy to reduce foreign dependencies. But they cannot keep doing more of all these things without limits or without tradeoffs among different uses and values.

Choices must be made about which uses and values to favor and how to balance management to get the best mix from a limited land base (Niemi, Mendelsohn, and Whitelaw 1991). National forests and grasslands hold large market shares of many valuable resources, some of which may be available only from these lands in the future. And they provide vital environmental services, some of which may not be provided by other land ownerships.

Because the United States is such a large force in global resource consumption and its environmental ramifications, choices made for national forests and grasslands should consider what these lands are best capable of providing and what will likely happen on private lands and the forests and rangelands of other nations as a result of the choices made for national forests and grasslands.

The essential policy question is a variation on that posed by Clawson (1975): "Given the capabilities of national forests and grasslands relative to other wildlands, and given the needs and wants of Americans for various economic and environmental benefits, what are the desired current and future conditions, uses, and values of our national forests and national grasslands?" Answers to this question will lead to further questions, such as How should these conditions, uses, and values be restored, created, or maintained? and How should those who most directly benefit from these conditions and roles pay for the benefits or compensate those who are not benefited?

THE MOVE TO ECOSYSTEM MANAGEMENT

Programs and plans for the National Forest System are where such questions get answered. The answers change periodically to reflect new information and changing social and economic circumstances.

Strategic direction for managing the National Forest System is set for five-year periods under the Resources Planning Act (RPA) of 1974. The current version, the 1990 RPA Program, identifies four themes and nineteen contemporary issues for attention. The balance of management investments among the various multiple uses is being improved through increased attention to recreation, wildlife, and fisheries resources. Commodity production programs are being examined and adjusted when necessary to ensure that they conform to environmental protection standards. Research on natural resources and how ecological systems function is being increased. And research, resource management, technical assistance, and international programs are addressing global resource issues. Each of these themes represents a commitment to sustain the long-term health, diversity, and productivity of the land and provide for wise use of the land's many resources to serve people.

Principal forestry issues receiving attention in and beyond the National Forest System in the 1990 RPA Program include the expansion of U.S. roles in global environmental stewardship, conservation of biological diversity, protection of endangered species, riparian area management, improvement of range condition, maintenance of water and air quality, control of catastrophic fires, conservation of old-growth forests, recovery of spotted owls, reduction of below-cost timber sales, reduction of clearcutting, increase in the near-term softwood timber supply, enhancement of timber supply from nonindustrial private lands, improvement of wilderness management, response to changing recreation needs, environmentally sensitive minerals development, financing management options, and reduction of appeals and litigation over forest management decisions.

Each national forest has an integrated land and resource management plan that, as appropriate to the forest, addresses these and other local issues and needs. The plans are developed and kept current through an open, public involvement process, often in collaboration with conservation partners in other governmental agencies and the private sector. The plans guide overall management of all of the lands and resources in the various units of the National Forest System.

NEW PERSPECTIVES PROJECTS: LEARNING HOW TO SUSTAIN ECOSYSTEM CONDITIONS FOR BROADER BENEFITS

To carry out the management directions of the RPA Program, forest plans, and annual budgets, teams of land managers, scientists, academicians, and citizens develop on-the-ground projects. Some of these projects were designed to test the practical application of emerging scientific information, new technologies, and new partnerships with the public. The special agency program called New Perspectives heightened focus on these integrated projects during 1990-92.

There were five primary goals for New Perspectives projects (Salwasser 1991a): (1) to improve understanding of how to sustain ecological systems at multiple geographic scales for a richer variety of current and future benefits and uses, (2) to better integrate all aspects of land and resources management, (3) to strengthen partnerships between forest managers and forest users, (4) to improve the effectiveness of public participation in resource decision making, and (5) to strengthen teamwork between researchers and resource managers in carrying out adaptive land and resource management.

New Perspectives projects were used to shape an ecosystem management perspective that is also emerging in other nations (Franklin et al. 1989; Plochmann 1989; Maini 1990). The Swedish, for example, call their approach *Rikare Skog*, a richer forest (Skogsstyrelsen 1990). A richer forest means a wider variety of values, uses, and services from current and future forest ecosystems. It means designing forest management with diversity and sustainability of the entire ecosystem in mind (Society of American Foresters 1991; Keystone Center 1991; Hansen et al. 1991; Salwasser 1991b; Reid et al. 1992). For further discussion, it's important to gain a common understanding of what we mean by the terms *ecosystem, landscape*, and *biodiversity*.

Ecosystems are communities of organisms working together with their environments as integrated units. They are places where all plants, animals, soils, waters, climate, people, and processes of life interact as a whole. These ecosystems/places may be small, such as a rotting log, or large, such as a continent or the biosphere. The smaller ecosystems are subsets of the larger ecosystems; that is, a pond is a subset of a watershed, which is a subset of a landscape, and so forth. All ecosystems have flows of things—organisms, energy, water, air, nutrients—moving among them. And all ecosystems change over space and time (Thomas 1956; Burgess and Sharpe 1981; Shugart 1984; Waring and Schlesinger 1985; Botkin 1990; Kimmins 1992). Therefore, it is not possible to draw a line around an ecosystem and mandate that it stay the same or stay in place for all time. Managing ecosystems means working with the processes that cause them to vary and to change.

Landscapes are relatively large areas that have similar and repeatable patterns of physical features, habitats, and human communities. They are the geographic context for planning the management of ecosystems. A good way to think about landscapes is that they are what you see when you stand on a vista point and look out over the land.

Biological diversity is the variety of life in an area (i.e., genes, species, populations of species, the symbiotic associations of species that ecologists call *biological communities*, and the many processes through which all of the biological parts of ecosystems are interconnected with all of the physical parts through space and time). Biological diversity at larger geographic scales, such as watersheds, landscapes, and beyond, includes the diversity of human cultures and lifestyles.

PRINCIPLES FOR ECOSYSTEM MANAGERS

"Ecosystem management means using an ecological approach to achieve the multiple-use management of national forests and grasslands by blending the needs of people and environmental values in such a way that national forests and grasslands represent diverse, healthy, productive, and sustainable ecosystems" (Robertson 1992). It is more a process and way of thinking than it is a rigid set of practices. Therefore, principles for guiding locally adapted management goals and practices are more useful than specific "top-down" instructions to "do the same thing everywhere."

There can be many different goals or biological conditions toward which ecosystem management can strive. Different biological and physical capabilities of ecosystems will define what is possible and their likely responses to management. When these potentials and responses are combined with people's needs and economic feasibility, specific goals for managing particular ecosystems will begin to emerge.

The Forest Service considers open public participation, partnerships with other public and private groups, and strong teamwork between scientists and managers to be critical to the eventual success of ecosystem management. It also encourages the practical use of four principles to guide the evolution of ecosystem management (Robertson 1992). These principles, stated in current terminology, derive from the agency's long-standing legal mandates.

- Protect land health by restoring or sustaining the integrity of soils, air, waters, biological diversity, and ecological processes, thereby sustaining what Aldo Leopold (1949) called the *land community* and what we now call *ecosystems*.
- Within the sustainable capability of the land, meet the needs of people who depend on natural resources for food, fuel, shelter, livelihood, and inspirational experiences.
- Contribute to the social and economic well-being of communities,

regions, and the nation through cost-effective and environmentally sensitive production and conservation of natural resources such as wood, water, minerals, energy, forage for domestic animals, and recreation opportunities, again within the sustainable capability of the land.

• Seek balance and harmony between people, land, and resources with equity between interests, across regions, and through generations, meeting this generation's resource needs while maintaining options for future generations also to meet their needs.

The World Commission on Environment and Development (1987) called its version of these principles *sustainable development.* They are consistent in spirit with principles from the United Nations Conference on Environment and Development (United Nations 1992).

THE IMPLEMENTATION OF ECOSYSTEM MANAGEMENT

Foresters, rangeland managers, fisheries managers, and wildlife habitat managers have historically managed ecosystems for certain desired uses and conditions. For example, one traditional goal of management was to produce and sustain the yields of selected products, such as wood, wood fiber, livestock forage, game wildlife, water, fish, or recreation. Management to maximize or optimize a single or selected few products of ecosystems usually involves simplifying the system, for example, through clearcutting followed by planting desired tree species and reducing competing vegetation. (There is a counterpart to this form of resource management for range, fish, and wildlife resources.)

An ecosystem perspective seeks to sustain diversity as well as productivity and to fit management practices at different geographic scales to suit best the characteristics of the land and the specific purposes for which different areas are being managed. This does not *a priori* exclude any form of resource production or nature preservation as long as it contributes to overall goals without degrading the long-term diversity and productivity of the system. It does, however, stress harmonizing the protection of natural systems with the production of resource uses commensurate with protection. In most cases the management of ecosystems on public wildlands will strive for multiple benefits rather than single-resource emphasis.

BIODIVERSITY GUIDELINES

Conserving biological diversity and productivity is central to an ecosystem perspective on land and resource management. It requires specific, measurable, achievable, and consistent objectives for what is going to be conserved, in what conditions, and in what places (Salwasser 1991c). A strategic framework for conserving diversity on U.S. federal lands was developed through national policy dialogue during the late 1980s (Keystone Center 1991). This framework has been adapted for Forest Service ecosystem management.

Where appropriate, objectives for biodiversity should be represented in regional guides and forest plans in the form of specific land-use allocations, management standards, or working guidelines for areas of lands or waters. These objectives can also serve as indicators of how well a diversity conservation strategy is working.

Recover and Conserve Formally Listed Threatened or Endangered Species

Listed species are the most vulnerable officially recognized elements of biodiversity. Specific objectives should lead to protection of existing populations and habitats of listed species and their restoration to viable levels if necessary. A net decline in the number of listed species in the area covered by a management plan or program is the desired end result.

Provide for Viable Populations of Native Plant and Animal Species

Species whose demographic or habitat trends are negative but not yet to the point of endangerment are the next most vulnerable elements of biological diversity. Some such species may even be more vulnerable than officially listed species. Habitats, human activities, and the demographics of wild populations of plants and animals should be managed to assure that populations of native species are viable and well distributed throughout their geographic ranges (Salwasser 1988). This requires a combination of actions to protect, restore, and enhance sufficient kinds, amounts, qualities, and distributions of subpopulations and habitats. The desired result is to secure the places and functions of all native species in regional ecosystems before they reach the point where formal listing as a threatened or endangered species calls into play the extreme measures of protecting species under crisis conditions (Salwasser 1991a). Especially important in achieving population viability is the perpetuation of multiple, interconnected, demographically resilient local populations; the characteristic genetic variation of the entire species; and the full range of the species' roles in ecological processes. Principles of conservation biology (Soulé and Wilcox 1980; Soulé 1986, 1987) and especially the population viability analysis and management process described by Marcot, Holthausen, and Salwasser (in press) are useful in this task.

Maintain a Viable Network of Native Biological Communities and Ecosystems

Certain biological communities or successional stages (often highly productive sites, riparian areas, and mature or old-growth successional stages) are likely to be vulnerable elements of biological diversity in managed landscapes. Lands, human activities, and habitats should be managed to assure that a network of representative native biological communities is maintained across the landscape (Harris 1984; Scott et al. 1987; Scott, Csuti, and Davis 1991; Keystone Center 1991). This may involve ecological restoration in some cases. Especially important are communities or assemblages of species that are rare or imperiled in the region or nation (Jenkins 1988). The "matrix conditions" of landscapes should also provide essential resources for all species to the degree that this is possible, including conditions needed for natural movements of plants and animals throughout the landscape and for the full range of ecological processes characteristic to the area. Where this is not possible, a specific network of reserves and connections between them may be needed (Harris and Gallagher 1989). The reserves, landscape matrix, and connections must be sufficiently large and diverse to accomplish their intended purposes.

Maintain Structural Diversity

Natural elements of structural diversity such as snags, caves, fallen trees, and seeps provide habitats for many species that would not live in an area without them. These elements can be jeopardized by intensive human uses such as fuelwood gathering and water diversions. They should be maintained in qualities, amounts, and distributions within patches and across landscapes to assure their roles in sustaining the diversity, productivity, and resilience of ecosystems from site to regional geographic scales (Franklin 1988).

Sustain Genetic Diversity

The genetic variation of intensively managed wild plant and animal populations can decline if sufficient attention is not paid to the effects of human selection for various traits. Populations should be managed to sustain natural levels of genetic variation within and among populations and the genetic integrity of representative and extreme populations (Ledig 1986; Millar 1987).

Produce and Conserve Resources Needed by People

People will obtain natural resources from somewhere. The key is to produce them in ways that do not lead to undesired environmental effects at local, regional, or global scales. If resources can be produced in ways that reduce human pressures on biological diversity in other places, then resource production zones can have a positive overall effect on biodiversity conservation. High-productivity sites, such as flat ground with deep loamy soils, and featured species, such as pines, firs, oaks, elk, and trout, should be managed with state-of-the-art efficiency in certain places to sustain the production of resources desired by people, thus meeting human needs with minimal effects on more fragile sites and sensitive species.

Protect Ecosystem Integrity—Soils, Waters, Biota, and Ecological Processes

Every human activity has some effect on lands, waters, or biota. Ideally, these effects can be minimized through sensitivity to ecosystem integrity. Actions that are known to degrade site conditions or long-term ecosystem diversity, productivity, or resilience should be avoided if possible or mitigated promptly when not. Resource management activities should use the natural restorative powers of ecosystems to the maximum extent feasible. The kinds, amounts, and distribution of living and dead organic matter to be left in the ecosystem for long-term renewal of the land after resource harvest should be routinely considered along with how much of the productivity of the system is to be removed for human uses.

Restore and Renew Degraded Ecosystems

Biological communities, waters, and soils that have been damaged by natural events or past human actions should be under restoration and renewal programs that embrace the concepts and methods of restoration ecology and management (Bonnicksen 1988; Cairns 1986, 1988a, 1988b, 1988c, 1989; Jordan, Gilpin, and Aber 1987; Jordan 1988).

LANDSCAPE AND SITE DESIGN DIFFERENCES

Ecosystem management aims for a richer diversity of resource uses and values and of native plant and animal species, biological communities, and ecological processes than was provided by historic multiple-use management. But managing ecosystems to sustain a richer set of benefits and future options does not mean that all sites will receive the same treatment or serve identical purposes (Forman and Godron 1986; Hunter 1990; Kimmins 1992). Because each site can potentially serve different purposes, the design task is to determine what is needed to meet overall goals; design compatible blends of conditions, uses, and values for different ecosystems; and then assign sites and watersheds to the "package" of desired future conditions that best suit overall goals. The product of this design task is a map that shows a mosaic of different land-use classes that will best restore or sustain desired ecosystem conditions and provide for desired resource values, uses, and services (Caplan 1992). This, of course, is not a new idea, but it does require changes in how single-resource managers conduct their business.

At the landscape scale, the management of national forests and grasslands with an ecosystem perspective will require a greater variety of land-use classes than was historically used to achieve desired conditions, values, and uses. For example, areas designated in forest plans for the primary purpose of restoring and protecting native ecosystems and rare elements of biological diversity may be under several different kinds of management direction to accomplish specific goals. Some of these areas have been designated by the Congress, such as wilderness areas and wild and scenic rivers. Others, such as research natural areas, are designated by the chief of the Forest Service. In these areas, natural processes are encouraged, but some human intervention may be necessary to sustain desired ecological conditions (e.g., prescribed fires).

Areas designated in forest plans for multiple benefits also will have a wider variety of management direction under an ecosystem perspective. For example, multiple-benefit areas in forest plans may have site-scale management direction to sustain substantial amounts and distributions of large live trees, standing and fallen dead trees, native hardwoods, riparian areas, the complex flora and fauna of the soil, and the seeds of diversity from native forests, what Franklin et al. (1989) call *biological legacies*. These legacies contribute to the long-term diversity, productivity, and resilience of ecosystems (Hansen et al. 1991; Swanson and Berg 1991). A recent policy change by the Forest Service also limits the use of clear-cutting in multiple-use areas to better sustain biologically diverse and aesthetically pleasing forests (Robertson 1992).

Areas in the National Forest System that are designated for economically efficient production of wood, energy, minerals, water, recreation, and fiber to help serve the nation's needs for resources will have stronger environmental protection measures under ecosystem management. These areas may still be dominated by a single or few resource goals. Even these intensively managed wildlands, however, will feature the protection of riparian areas, within-stand species diversity, contributions to carbon sequestration, habitat for early successional wildlife, and outdoor recreation.

Public Participation: Choosing Desired Present and Future Conditions

Following the principles and guidelines for ecosystem management cited in this chapter, choices on what to do about U.S. national forests and grasslands should address desired present and future conditions of three things: environments, economies, and communities (recall figure 2.4). The choices should consider potential and desired conditions at several geographic scales, at least at local, regional, and national scales and at global scales if possible and feasible. And the choices should consider at least the next several decades, if not the next several generations.

Because such choices are inherently value laden and political, public involvement in public resource decision making must become more effective. Managers and scientists cannot make these choices for people, but they can and should inform people about the potentials and likely consequences of options. Ecosystem management, thus, entails new forms of public participation (Wondolleck 1988) and new roles for managers and scientists in forest policy.

CONCLUSION

Debates over the sustainability of forests and other natural resources are often tinged with overtones of despair and even eminent catastrophe. Calls have been made for wholesale changes in institutions and societal priorities to address the "crisis" situation. Debate is healthy in a democratic society, but it should be based on facts and objective portrayals of potentials and options. Furthermore, before people decide where they want to go, they might consider where they have been, where they are, and how they got there.

To address resource depletion in the late nineteenth century, conservation was offered as a model for ethical behavior regarding forests and forest resources. It stressed protection of soils, waters, and biota; scientific management; and the wise use of resources to serve people's needs. Over the years, new dimensions emerged, including multiple use, sustained yield, wilderness preservation, endangered species protection, and integrated land management and planning.

There is overwhelming evidence that, while some problems remain and others are emerging, on balance, conservation and multiple use have served Americans well. Conditions of U.S. forests, wildlife, rangelands, agricultural lands, and related resources may not meet the aspirations of all citizens, but they have improved dramatically during the last century. These trends continue. The current resource situation in the United States offers a much broader range of choices than would have existed had conservation and multiple-use policies not been put into place. As we consider changes in forest policy direction for the future, this historical dimension should not be forgotten.

Although conservation and multiple-use management are still good models for the future, they are rightly taking on new dimensions. The challenges in sustaining diverse ecosystems in the face of a growing human population call for renewed vigor in pursuit of the ideals of conservation, land stewardship, and multiple-use management (Robertson 1991). They also call for renewing individual responsibility and the sense of community among the people, land, and resources (Leopold 1949; Sirmon 1991).

Managing ecosystems to conserve diverse, resilient, and productive forests and rangelands is conceptually broader than the previous models of sustained-yield resource management and multiple use, but it does not reject the accomplishments, contributions, or future utility of those concepts. In fact, it builds directly on a foundation established by previous policies, concepts, and accomplishments; it could not happen without them. It goes beyond them, however, and requires some changes in how we think and how we organize to accomplish work.

Because of what prior generations of political leaders, citizens, scientists, and resource managers created with a National Forest System and scientifically sound management, we can explore management models that consider more than selected resource outputs, more than single species, more than "island" parks and refuges, and more than a mechanistic, reductionist view of nature. Our goals can be broader because we paid the price to restore and maintain options.

The fact that so many forest resource options exist for our relatively affluent society probably means that controversy over forest uses in the United States will not soon subside. Other models will be offered for how to conserve national forests and national grasslands. Some will favor more preservation of land for the conditions and values that natural events will produce, up to 50 percent of the total area of the United States in wilderness in one case (Noss 1991). Others will favor more development of natural resources to serve immediate human needs. Conflict over uses and values will probably not decline much until people reach accord on the basic policy question: What roles, values, and uses are desired for public lands?

Ecosystem perspectives now taking shape across North American forests offer ecologically based management and complementarity of sites—nature preserves, resource conservation areas, resource production areas, and recreation areas—as a model for designing landscape management to protect environmental values while still providing for people's needs, especially those of local people. Regardless of the specific dimensions of future management of national forests and grasslands, it will proceed without perfect information (Holling 1978) and people must be ready for change. Forest ecosystems will change whether humans want them to or not (Kimmins 1992). Furthermore, it will not be possible to sustain desired forest conditions in a changing world without a reasonable degree of human well-being and vice versa. Thus, education, economic development, equity in the distribution of resources, and conservation of natural resources—all to improve the lives of people—must complement good land management as the necessary and sufficient parts of a globally responsible land ethic (Reid et al. 1992; United Nations 1992). Remember what Garrett Hardin said, "it is not possible to do only one thing in an ecosystem" (Hardin 1985).

REFERENCES

Alexander, S., and B. Greber. 1991. Environmental ramifications of various materials used in construction and manufacture in the United States. Gen. Tech. Rep. PNW-GTR-277. Portland, OR: USDA Forest Service, Pacific Northwest Research Station.

Allen, T. F. H., and T. B. Starr. 1982. *Hierarchy: Perspectives for ecological complexity.* Chicago: University of Chicago Press.

Baden, J. 1991. Spare that tree! *Forbes,* Dec. 9:229-233.

Bingham, C. W. 1991. Forest resource availability and use: Wood and timber from a U.S. perspective. Presented at the Society of American Foresters National Convention, San Francisco, August.

Binkley, C. S. 1991. The global economy and rising expectations. Presented at the Wood Product Demand and the Environment Conference, Forest Products Research Society, Vancouver, Canada, November 13-15.

Bonnicksen, T. M. 1988. Restoration ecology: Philosophy, goals, and ethics. *Environmental Professional* 10:25-35.

Botkin, D. B. 1990. *Discordant harmonies: A new ecology for the twenty-first century.* New York: Oxford University Press.

Bowyer, J. L. 1991a. Responsible environmentalism—the ethical features of forest harvest and wood use. Presented at the National Conference on Ethics in America, Long Beach, CA.

————. 1991b. Responsible environmentalism: The ethical features of forest harvest and wood use on a global scale. *Forest Perspectives* 1(4): 12-14.

————. 1991c. Resource management: A need for realistic assumptions, global thinking. Presented at the National Stewardship Conference, Duluth.

Brown, L. R. 1992. *State of the world.* New York: W.W. Norton & Co.

Burgess, R. L., and D. M. Sharpe, eds. 1981. *Forest island dynamics in man-dominated landscapes.* New York: Springer-Verlag.

Cairns, J., Jr. 1986. Restoration, reclamation, and regeneration of degraded or destroyed habitats. In *Conservation biology: The science of scarcity and diversity,* ed. M. E. Soulé, 465-484. Sunderland, MA: Sinauer Associates.

————, ed. 1988a. *Rehabilitating damaged ecosystems.* 2 vols. Boca Raton, FL: CRC Press.

————. 1988b. Restoration ecology: The new frontier. In *Rehabilitating damaged ecosystems,* ed. J. Cairns, Jr., 1:1-11. Boca Raton, FL: CRC Press.

————. 1988c. Restoration of damaged ecosystems and opportunities for increasing diversity. In *Biodiversity,* ed. E. O. Wilson, 333-343. Washington, DC: National Academy Press.

————. 1989. Restoring damaged ecosystems: Is predisturbance condition a viable option? *Environmental Professional* 11:152-159.

Caplan, J. A. 1992. Striding into elephant country. *George Wright Forum* 9(4).

Caulfield, C. 1990. The ancient forest. *New Yorker*, May 14:46-84.

Chappelle, D. E., and H. H. Webster. 1991. Natural resources and societal prosperity: Linkages, opportunities and dangers. Presented at the Wood Product Demand and the Environment Conference, Forest Products Research Society, Vancouver, November 13-15.

Clark, R. N., and G. H. Stankey. 1991. New forestry or new perspectives? The importance of asking the right question. *Forest Perspectives* 1(1):9-13.

Clawson, M. 1975. *Forests for whom and for what?* Baltimore: Johns Hopkins University Press.

————. 1979. Forests in the long sweep of American history. *Science* 204:1168-1174.

CORRIM (Committee on Renewable Resources for Industrial Materials—National Research Council). 1976. Wood for structural and architectural purposes. *Wood and Fiber* 8(1):1-72.

Dunlap, T. R. 1989. *Saving America's wildlife*. Princeton: Princeton University Press.

Ehrlich, P. 1990. *Population explosion*. New York: Simon & Schuster.

Fedkiw, J. 1989. The evolving use and management of the nation's forests, grasslands, croplands, and related resources. Gen. Tech. Rep. RM-175. Fort Collins, CO: USDA Forest Service, Rocky Mountain Forest and Range Experiment Station.

Forman, R. T. T., and M. Godron. 1986. *Landscape ecology*. New York: John Wiley & Sons.

Franklin, J. F. 1988. Structural and functional diversity in temperate forests. In *Biodiversity*, ed. E. O. Wilson, 166-175. Washington, DC: National Academy Press.

Franklin, J. F., D. A. Perry, T. D. Schowalter, M. E. Harmon, A. McKee, and T. A. Spies. 1989. Importance of ecological diversity in maintaining long-term site productivity. In *Maintaining the long-term productivity of Pacific Northwest forest ecosystems*, eds. D. A. Perry, B. Thomas, and R. Meurise, 82-97. Portland, OR: Timber Press.

Frederick, K. D., and R. A. Sedjo, eds. 1991. *America's renewable resources: Historical trends and current challenges*. Washington, DC: Resources for the Future.

Fri, R. W. 1991. Sustainable development: Can we put these principles into practice? *Journal of Forestry* 89(7):24-25.

Gale, R. P., and S. M. Cordray. 1991. What should forests sustain? Eight answers. *Journal of Forestry* 89(5):31-36.

Gillis, A. M. 1990. The new forestry; an ecosystem approach to land management. *BioScience* 40(8):558-562.

Greber, B. J., and K. N. Johnson. 1991. What's all the debate about overcutting?
Journal of Forestry 89(11):25-30.

Hansen, A. J., T. A. Spies, F. J. Swanson, and J. L. Ohmann. 1991. Conserving
biodiversity in managed forests. *BioScience* 41(6):382-392.

Hardin, G. 1985. *Filters against folly: How to survive despite economists, ecologists, and the merely eloquent.* New York: Penguin Books.

Harris, L. D. 1984. *The fragmented forest: Island biogeography theory and the
preservation of biotic diversity.* Chicago: University of Chicago Press.

Harris, L. D., and P. B. Gallagher. 1989. New initiatives for wildlife conservation:
The need for movement corridors. In *Preserving communities and corridors,*
ed. G. Mackintosh, 11-34. Washington, DC: Defenders of Wildlife.

Haynes, R. W. 1990. An analysis of the timber situation in the United States:
1989-2040. Gen. Tech. Rep. RM-199. Fort Collins, CO: USDA Forest Service,
Rocky Mountain Forest and Range Experiment Station.

Haynes, R. W., and D. J. Brooks. 1991. Wood and timber availability from a
Pacific rim perspective. In *Proceedings of the Annual Convention of the Society
of American Foresters, San Francisco.*

Holling, C. S., ed. 1978. *Adaptive environmental assessment and management.*
New York: John Wiley & Sons.

Hunter, M. L., Jr. 1990. *Wildlife, forests, and forestry.* Englewood Cliffs, NJ:
Prentice-Hall.

Ince, P. J., and J. T. Alig. 1991. Wastepaper recycling and the future timber
market. Presented at the USDA Annual Outlook Conference, December 2.

Jenkins, R. E., Jr. 1988. Information management for the conservation of biodiversity. In *Biodiversity,* ed. E. O. Wilson, 231-239. Washington, DC: National Academy Press.

Johnson, K. N., J. F. Franklin, J. W. Thomas, and J. Gordon. 1991. Alternatives for
management of late-successional forests in the Pacific Northwest: A report to
the U.S. House of Representatives. Washington, DC: Committee on Agriculture, Subcommittee on Forests, Family Farms, and Energy, and Committee on
Merchant Marine and Fisheries, Subcommittee on Fisheries and Wildlife,
Conservation, and the Environment.

Jordan, W. R., III. 1988. Ecological restoration: Reflections on a half-century of
experience at the University of Wisconsin-Madison Arboretum. In *Biodiversity,* ed. E. O. Wilson, 311-316. Washington, DC: National Academy Press.

Jordan, W. R., III, M. E. Gilpin, and J. D. Aber, eds. 1987. *Restoration ecology.*
Cambridge: Cambridge University Press.

Keystone Center. 1991. Final consensus report of the Keystone Policy Dialogue
on Biological Diversity on Federal Lands. Keystone, CO.

Kimmins, H. 1992. *Balancing act: Environmental issues in forestry.* Vancouver: UBC Press.

Knize, P. 1991. The mismanagement of the national forests. *Atlantic* 268(4):98-112.

Knudsen, T. 1991. The Sierra in peril: Special report. *Sacramento Bee*, Sacramento, CA.

Koch, P. 1991. Wood vs non-wood materials in U.S. residential construction: Some energy-related international implications. Working Paper 36. Seattle: Center for International Trade in Forest Products, University of Washington.

Kohm, K. A. 1991. *Balancing on the brink of extinction: The Endangered Species Act and lessons for the future.* Washington, DC: Island Press.

Lal, J. B. 1989. *India's forests: Myths and realities.* Dehra Dun, India: Nostraj Publishers.

Ledig, F. T. 1986. Heterozygosity, heterosis, and fitness in outbreeding plants. In *Conservation biology: The science of scarcity and diversity*, ed. M. E. Soulé, 77-109. Sunderland, MA: Sinauer Associates.

Leopold, A. 1949. *A Sand County almanac and sketches here and there.* New York: Oxford University Press.

Lubchenko, J., A. M. Olson, L. B. Brubaker, S. R. Carpenter, M. M. Holland, S. P. Hubbell, S. A. Levin, J. A. MacMahon, P. A. Matson, J. M. Melillo, H. A. Mooney, C. H. Peterson, H. R. Pulliam, L. A. Deal, P. J. Regal, and P. G. Risser. 1991. The sustainable biosphere initiative: An ecological research agenda. *Ecology* 72(2):371-412.

MacCleery, D. W. 1991. *Condition and trends of U.S. forests: A brief overview.* Washington, DC: USDA Forest Service, Timber Management Staff.

Maini, J. S. 1990. Sustainable development and the Canadian forest sector. *Forestry Chronicle*, Aug.:346-349.

Marcot, B. G., R. S. Holthausen, and H. Salwasser. A process for population viability assessment and planning. In *Viable populations*, ed. B. G. Marcot and D. Murphy. New York: Oxford University Press. In press.

Marsh, G. P. 1864. *Man and nature; or, physical geography as modified by human actions.* New York: Scribners.

McKibben, W. 1989. *The end of nature.* New York: Bantam Books.

Millar, C. 1987. The California forest germplasm conservation project: A case for genetic conservation of temperate tree species. *Conservation Biology* 1:191-193.

Mitchell, J. G. 1990. *War in the woods*, 84-121. New York: Audubon.

National Research Council. 1990. *Forestry research: A mandate for change.* Washington, DC: National Academy Press.

Niemi, E., R. Mendelsohn, and E. Whitelaw. 1991. New conflicts stir managers of U.S. forests. *Forum for Applied Research and Public Policy* 6(3):5-12.

Noss, R. F. 1991. Wilderness and sustainability. *Conservation Biology* 5(1):120-122.

Nyland, R. D. 1991. Exploitation and greed in eastern hardwood forests. *Journal of Forestry* 90(1):33-37.

O'Toole, R. 1988. *Reforming the Forest Service*. Washington, DC: Island Press.

Perlin, J. 1991. *A forest journey: The role of wood in the development of civilization*. Cambridge: Harvard University Press.

Plochmann, R. 1989. The forest of Central Europe: A changing view. Presented at the Starker Lectures, Oregon State University, Corvallis, OR.

Postel, S., and J. C. Ryan. 1991. Reforming forestry. In *State of the world 1991: A Worldwatch Institute report on progress toward a sustainable society*, ed. L. Starke, 74-92. New York: W.W. Norton & Co.

Reid, W., C. Barber, and K. Miller. 1992. *Global biodiversity strategy: Guidelines for action to save, study, and use earth's biotic wealth sustainably and equitably*, ed. K. Courrier. Washington, DC: World Resources Institute (WRI), The World Conservation Union (IUCN), and United Nations Environment Programme (UNEP) in cooperation with the Food and Agriculture Organization (FAO) and United Nations Education, Scientific and Cultural Organization (UNESCO).

Robertson, F. D. 1991. The next 100 years of national forest management. *Transactions of the North American Wildlife and Natural Resources Conference* 56:19-21.

————. 1992. Ecosystem management of the national forests and grasslands. Memo to Regional Foresters and Station Directors, USDA Forest Service, Washington, DC: June 4.

Salwasser, H. 1988. Managing ecosystems for viable populations of vertebrates: A focus for biodiversity. In *Ecosystem management for parks and wilderness*, eds. J. K. Agee and D. R. Johnson, 87-104. Seattle: University of Washington Press.

————. 1991a. New perspectives for sustaining diversity in the U.S. National Forest System. *Conservation Biology* 5(4):567-569.

————. 1991b. In search of an ecosystem approach to endangered species conservation. In *Balancing on the brink of extinction: The Endangered Species Act and lessons for the future*, ed. K. A. Kohm, 247-265. Washington, DC: Island Press.

————. 1991c. Roles for land managers in conserving biological diversity. In *Challenges in the conservation of biological resources*, eds. D. J. Decker, M. E. Krasny, G. R. Goff, C. R. Smith, and D. W. Gross, 11-32. Boulder, CO: Westview Press.

Sample, V. A. 1991. Bridging resource use and sustainability: Evolving concepts of both conservation and forest resource management. Presented at the Society of American Foresters National Convention, San Francisco, August.

Scott, J. M., B. Csuti, and F. Davis. 1991. Gap analysis: An application of geographic information systems for wildlife species. In *Challenges in the conservation of biological resources: A practitioner's guide*, eds. D. J. Decker, M. E. Krasny, G. R. Goff, C. R. Smith, and D. W. Gross, 167-180. San Francisco: Westview Press.

Scott, J. M., B. Csuti, J. D. Jacobi, and J. E. Estes. 1987. Species richness: A geographic approach to protecting future biological diversity. *BioScience* 37(11):782-788.

Sedjo, R. A. 1990. *The national forest resources*. Discussion paper ENR90–07. Washington, DC: Resources for the Future.

Sedjo, R. A., and K. S. Lyon. 1990. *The long-term adequacy of world timber supply*. Washington, DC: Resources for the Future.

Shands, W. E. 1991. *The lands nobody wanted: The legacy of the eastern national forests*. Milford, PA: Pinchot Institute for Conservation.

Shugart, H. H. 1984. *A theory of forest dynamics: The ecological implications of forest succession models*. New York: Springer-Verlag.

Silver, C. S., and R. S. DeFries, eds. 1990. *One earth, one future: Our changing global environment*. Washington, DC: National Academy Press.

Sirmon, J. M. 1991. A new ideal of leadership in natural resources. Presented at the National Forest Centennial Futures Conference, Atlanta, April.

Skogsstyrelsen. 1990. *Rikare skog: 90-Talets Kunskaper om Naturvard och Ekologi*. Jonkoping, Sweden: Skogsstyrelsen.

Society of American Foresters. 1991. *Biological diversity in forested ecosystems: A position statement of the Society of American Foresters*. Bethesda, MD.

Soulé, M. E. 1986. Conservation biology and the real world. In *Conservation biology: The science of scarcity and diversity*, ed. M. E. Soulé, 1-12. Sunderland, MA: Sinauer Associates.

————, ed. 1987. *Viable populations for conservation*. New York: Cambridge University Press.

Soulé, M. E., and B. A. Wilcox, eds. 1980. *Conservation biology: An evolutionary-ecological perspective*. Sunderland, MA: Sinauer Associates.

Swanson, F., and D. Berg. 1991. The ecological roots of new approaches to forestry. *Forest Perspectives* 1(3):6-8.

Terborgh, J. 1990. *Where have all the birds gone?* Princeton: Princeton University Press.

Thomas, J. W. 1989. Wildlife resources. In *Natural resources in the 21st century*, eds. R. N. Sampson and D. Hair, 175-204. Washington, DC: American Forestry Association and Island Press.

Thomas, J. W., E. D. Forsman, J. B. Lint, E. C. Meslow, B. R. Noon, and J. Verner. 1990. *A conservation strategy for the northern spotted owl: Interagency*

Scientific Committee to address the conservation of the northern spotted owl.
Portland, OR: US Government Printing Office.

Thomas, W. L., Jr., ed. 1956. *Man's role in changing the face of the earth: An international symposium under the co-chairmanship of Carl O. Sauer, Marston Bates, and Lewis Mumford.* Chicago: University of Chicago Press.

Toynbee, A. 1976. *Mankind and mother earth: A narrative history of the world.* New York: Oxford University Press.

Trefethen, J. B. 1975. *An American crusade for wildlife.* New York: Winchester Press and the Boone and Crockett Club.

Ulrich, A. H. 1989. *U.S. timber production, trade, consumption, and price statistics 1950-87.* USDA Forest Service miscellaneous publication 1471. Washington, DC: USDA Forest Service.

_____. 1990. *U.S. timber production, trade, consumption, and price statistics 1960-88.* USDA Forest Service miscellaneous publication 1486. Washington, DC: USDA Forest Service.

United Nations. 1992. Adoption of agreements on environment and development. Presented at the United Nations Conference on Environment and Development, Rio de Janeiro, Brazil, June.

USDA Forest Service. 1973. The outlook for timber in the United States. Forest Resource Rep. 20. Washington, DC.

_____. 1982. An analysis of the timber situation in the United States, 1952-2030. Forest Resource Rep. 23. Washington, DC.

_____. 1990. *The Forest Service Program for forest and rangeland resources: A long-term strategic plan.* Washington, DC: USDA Forest Service.

Waring, R. H., and W. H. Schlesinger. 1985. *Forest ecosystems: Concepts and management.* New York: Academic Press.

Williams, M. 1989. *Americans and their forests: An historical geography.* New York: Cambridge University Press.

Wondolleck, J. M. 1988. *Public lands conflict and resolution: Managing national forest disputes.* New York: Plenum Press.

World Commission on Environment and Development. 1987. *Our common future.* New York: Oxford University Press.

World Resources Institute. 1990. *World resources 1990-91.* New York: Oxford University Press.

Zonneveld, I. S. 1990. Scope and concepts of landscape ecology as an emerging science. In *Changing landscapes: An ecological perspective*, eds. I. S. Zonneveld and R. T. T. Forman, 3-20. New York: Springer-Verlag.

3

Institutional Constraints on Sustainable Resource Use: Lessons from the Tropics Showing That Resource Overexploitation Is Not Just an Attitude Problem and Conservation Education Is Not Enough

George Honadle
Hidden Creek Farm and University of Minnesota

Long-term access to natural resources is a problem for humans because current and past human actions have depleted those resources. The natural resource crisis results from human behavioral excesses of reproduction and exploitation. Recognizing this places social and institutional factors at center stage in discussions of sustainable resource stocks.

These two types of factors are often reduced to a problem of inadequate knowledge—if only people knew what they were doing, they would stop! And this problem statement translates into "conservation education" as the key to the solution. But experience with international development efforts suggest that this approach is inadequate—the gap between knowledge and behavior is wide, and it is the behavior that is the key.

This chapter explores some of that experience and suggests an alternative to the knowledge hypothesis. The perspective is that people respond to incentives for behavior that improves their survival chances and that a combination of economic market forces and bureaucratic policies, rules, and structures creates those incentives. This is where attention must be focused if sustainable societies are to be built.

Three examples make the point. A Philippine fisherman is destroying the coral reef that provides habitat and cover for the fish he harvests. He knows that his success in the present is dooming the fishing of the future. But, as he informs us, "I need the money now; what is the alternative?" In the second case, an Indian forester is taking bribes from villagers to allow them to keep "their" trees that the law says belong to the state. He knows that this is hindering the performance of the forestry effort, but his pay is so low he cannot live on it. This is his major source of income. What is he to do? In the third case, a farmer in Ecuador wants to establish a claim to forest land. To do this, the law requires him to fell trees and clear the land. So, even if he would prefer to keep some of the land forested, he is forced to remove the cover. And these cases are little more than the tip of the iceberg.

Population growth, market dynamics, and institutional parameters have created situations where exploitative behavior is rewarded and sustaining behavior is discouraged, even when ideologies proclaim the opposite. To understand how this happens, one must examine the causes of tropical deforestation and then identify institutional factors entangled in those causes. This may provide some insights into pitfalls and possibilities awaiting the promotion of sustainable forestry in North America.

THE PROBLEM OF DEFORESTATION IN THE TROPICS

The causes of deforestation are twofold. First, there are fundamental, or indirect, causes operating worldwide. Second, there are direct causes emanating from the fundamental ones and operating locally. The importance of specific causes will vary through space and time, but an overview of both is needed as background to an examination of the institutional dimension.

CAUSAL FACTORS

A fundamental and well-recognized cause of tropical deforestation is the increase in human population in tropical areas. This growth pushes people into more and more marginal areas in the search for land for food production. The edges of the forests thus are nibbled away by small parties or leveled on a larger scale by public or private settlement schemes.

Further inroads occur in the search for household fuelwood and homestead construction materials. The need by ever-increasing population for cooking and heating fuel has led to a widely recognized fuelwood crisis in the Third World. This crisis involves both the rural folk and the larger and larger clusters of urban populations.

These urban centers also create demands for more sophisticated energy types than wood to burn. Hydroelectric projects to feed the energy needs of growing industries and cities can flood vast tracts of virgin forest in a short time.

A second fundamental cause of deforestation is the penetration of international markets into Third World economies. Demand for forest products and alternative uses of forested land exert a pull on Third World people and their resources. The global village may be in the making, but the global marketplace is already here.

There are numerous manifestations of this phenomenon. The developed world's demand for beef has resulted in the clearing of forests for cattle production as Third World entrepreneurs respond to opportunities to make quick gains. Latin America has especially felt this influence.

The North American demand for fruit or for cheap and tasty cooking oils in the fast food industry has led to the clearing of complex, multi-species hardwood forests and their replacement by monoculture plantations of oil palm or citrus. The Ivory Coast is one example of this phenomenon. Another is Malaysia, which in the last decade has established itself as both the major world palm oil supplier and the primary source of genetic improvement in oil palm.

The hardwoods themselves are also sought by the industrialized nations for their own purposes. The Japanese extraction of hardwood from the forests of Papua New Guinea provides one example, but the phenomenon is limited neither to one industrialized country nor to one area of the globe.

In some African countries the tobacco industry is the major earner of foreign exchange for the local economy. It is also a major user of wood fuel for the drying process. Thus, smoking is harmful to the health of the indigenous forests.

A fifth example of international market penetration that threatens the tropical forests is the trade in illicit drugs emanating from South America and from Asia's "golden triangle." Forest clearing for coca or poppy production is a significant contributor to deforestation in some countries.

The sixth market-based cause of deforestation involves the industrial demand for precious and valuable metals. Mining often results in the total alteration of the earth's surface at the point of extraction, as well as downstream pollution of rivers, wetlands, estuaries, and coastal resources such as mangrove forests. Human warfare, which can be the extreme manifestation of market penetration, often demolishes forests as nations or groups establish control over territory or resources.

The view above, of two primary causes—population growth and market penetration—resulting in nine major secondary factors leading to tropical deforestation is represented in table 3.1. This depiction is not meant to be totally comprehensive; rather, it is indicative of the relationships among major primary and secondary causes of tropical deforestation in the latter part of the twentieth century.

First, as the table suggests, it is possible to take aim at the primary causes—population growth and market penetration. Family planning attempts to limit the demand for children as a way to lessen the push factor. Likewise, international conventions, trade barriers, and education cam-

TABLE 3.1. **Relationship of Primary and Secondary Causes of Tropical Deforestation**

Primary Causes (Indirect)	*Secondary Factors (Direct)*
Human population increase	Forests cleared for food production and settlement
	Trees cut for household fuelwood and construction
	Hydroelectric projects flood forests to meet energy demand
International market penetration	Forests cleared for beef production
	Hardwoods extracted from forests
	Monoculture plantations replacing forest
	Wood used for tobacco drying
	Forests cleared for drug production or warfare
	Mining operations altering the landscape

paigns in First World markets can aim at reducing the demand that pulls the carpet out from under the tropical forest. In fact, if population growth and market penetration were both checked, most of the other problems would wither away.

But these are difficult problems to solve, especially when many donor investments (in health and marketing, for example) are actually reinforcing them. Moreover, there is a tradeoff between short-term and long-term success. Solving the population problem now would not help now—the effects would appear much later. Although reversing the market influence now might have some immediate impact, unless the population problem were solved the long run prognosis would still be bad. Thus, a complete solution will involve both primary causes.

Alternatively, it is possible to sight in on the secondary factors. Limiting funding for dams, resettlement schemes, or plantation forestry that obliterates natural forest is one tactic. Developing better stoves and wood fuel-consuming equipment or pursuing alternative energy paths is another approach. Requiring mining methods that limit surface disruption and downstream pollution addresses one factor. Improving agricultural technologies that raise the food output per hectare or redistributing land to ease the pressure on the forests addresses another one. Both supply and demand dimensions can be tackled at this level. In all cases, however, strengthening the capacities of research, regulatory, production, and training institutions will be needed. Moreover, the relative importance of each secondary factor must be empirically established for particular countries or ecological zones. When this is done, priority investment areas may emerge from the exercise.

Both primary and secondary causes share a common difficulty. Attempted solutions at both levels quickly encounter complications, and many of those complications result from institutional factors.

INSTITUTIONAL COMPLICATIONS

Institutional factors complicate attempts to deal with these causes at three levels. First is the international level. For example, the debt crisis adds another dimension to market penetration by raising the pressure to use the forest as a foreign exchange mine. Thus, two contradictory messages are sent across the international channels. One says "adopt policies of fiscal responsibility, tighten your belt, unleash the private sector, and pay back (or at least service) the debt." This means focus on the short-term structural adjustment. Another message says "focus on long-term sustainable development and do not deplete the stock of natural resources—don't mine the forests."

Thus, international development and conservation agencies are often at odds with international financial institutions. It is harder to protect wildlands when exchange rates are dropped, requiring more exports to earn equal amounts of foreign exchange, or when public bureaucracies are cut back, weakening the institutions responsible for protecting and managing the forest resource and regulating the activities of those determined to mine it. Contradictory international institutional agendas make it difficult to address some of the factors causing deforestation.

Second, at the community or subnational level, social institutions complicate the picture. Inequitable landholding intensifies resettlement and forest clearing by protecting estates for the few while pushing the many to the margins. Political systems that have been captured by the wealthy can be driving forces in forest destruction. Alliances between local politicians and international agencies or merchants sometimes result, making solutions more elusive.

Moreover, the general tendency over the last two decades to bypass subnational governments and emphasize the national level of public management has left a power vacuum at the subnational level in many countries. The present weakness of managerial capacity at this level makes local resource management more difficult.

Additionally, local land ownership and tenure systems may not give incentives for people to safeguard natural resources. Grazing land, forests, and water are often considered common property resources (CPRs), and there are often pressures for individuals to use the CPR for their own advantage in the short run while depleting the resource base over the long run.

During the 1980s this question of the commons received high visibility and extensive examination. Much effort has gone into identifying effective traditional systems for managing CPRs and understanding how to build systems to meet the needs of today. Although the jury is still out on how to do it, some lessons are emerging.

- Some traditional societies developed effective rules to govern behavior and regulate resource squandering, and where such societies and rules continue to exist and work, it is desirable to support continued application of their resource management systems.

- Tree tenure or guaranteed access to and use of forest resources through traditional management systems at grassroots social levels can help to provide incentives for sustainable silviculture and non-destructive extraction of forest resources.
- Organizing human effort at the local level can help to restore degraded environments and initiate effective small-scale resource management.
- Both population growth and market penetration can bring destructive pressures to bear on traditional rule making and resource management systems and render them inadequate to contend with the new circumstances.
- Neither privatized ownership nor public regulation, the two major responses to the deterioration of the commons, offers clear-cut solutions to the problem.
- Many of the situations described as common property problems were not—they were rather what remained after common property systems vanished, or they were other regimes.

Thus, the nature of the property may yield less insight for future action than will examinations of bureaucratic behavior, policy conflicts, and the lack of incentives for sustainable resource use.

THE ORGANIZATION OF PUBLIC FORESTRY IN THE TROPICS

Surprisingly little work has been done to map the organization of public sector forestry management in the tropics. Even when heroic attempts are made to collect and synthesize data on the natural resource base, such as the *World Resources 1986-91* series, discussion of institutions is limited to lists of international organizations or identification of the countries that have signed a treaty. But listing the treaties to which different countries are signatories is a far cry from mapping the institutional factors that guide local policies and behavior. Treaty commitments can be more rhetorical than substantive when resources are scarce and national political will is divided—implementing the provisions of the treaty requires adequate administrative capacity. Moreover, a signature on a treaty is often more of a downstream effect than an upstream influence on a situation.

Formal institutional configurations do not always mirror exactly the true distribution of power or decision making that dominates in a country or organization because a combination of historical, structural, and behavioral elements can cause a similar organization to function differently under different circumstances. Both formal and informal dimensions are discussed below.

FORESTRY MANAGEMENT ORGANIZATIONS

The formal organization of authority to determine priorities for forestry management is both an outcome and a cause. It is an artifact of political power struggles to determine who gets to control the resource and the various benefits it may produce. Thus, to judge the informal pressures encountered by an agency, it is necessary to know the history accompanying its birth. At the same time, a choice to use one organization as opposed to another is a cause—it partly determines who will get access to the resource base and consequently how it will be used in the future. Moreover, it directly creates barriers and opportunities that program managers will encounter. Thus, attempts to put the brakes on runaway deforestation should be based on an understanding of the pitfalls and potentials associated with the use of different organizations.

Public sector responsibility for forestry management and protection is commonly entrusted to one of four types of national agencies in Third World countries. They are a parastatal body, a department inside a larger ministry, a separate ministry of forestry or natural resources, or a ministry of natural resources and tourism. Each of these configurations makes a statement about the perception of the forests held by national leaders and about how access to forest resources will be controlled.

One way of organizing the public sector responsibility for forestry is to establish a parastatal body outside the normal machinery of government and give it a free hand in its operations. This has been done in many countries. Although it is not universally true, often the decision to do this reflects the distribution of power in the country. When a small minority (military, racial, or tribal) holds an inordinate amount of power, a parastatal body is often preferred as a way of isolating control of forestry resources from popular pressures and letting a small group quietly get away with extracting the wealth. Thus, the artifact reflects the distribution of power and it also partly determines the constraints and opportunities surrounding those who manage the resource.

Numerous countries in Africa and Latin America show this pattern of organizational choice and resource use. Although it cannot be contended that this is an ironclad relationship between an organizational mechanism and forestry management, it is corroborated both by my own personal experience and by discussions with numerous development observers. A general pattern seems to be emerging.

Another common approach to forestry is to subsume it as a department under a ministry of agriculture or mines. In such a setting, agricultural production or mineral extraction gets priority attention and forestry is often a second-class citizen in the battle for financial resources. The agricultural version of this configuration can, however, promote the integration of forestry concerns and tree crops into the agricultural agenda.

Multiple uses of wood products, greater distribution of the tree cover, access to extension facilities and services, and various other factors distinguish this approach from that of the parastatal body (in the case of agriculture). A totally exploitive approach characterizes the other case.

But again, a static view is not warranted. In many countries there has been a constant shifting of the forestry oversight role from the ministry of agriculture to a separate ministry concerned, often, with forests and fisheries and then back again to agriculture. This history must be understood and placed within the context of the rise and fall of cabinet ministers, population growth, major sources of foreign exchange, the geographic distribution of forest wealth in relation to sources of political opposition to the dominant regime, and other historically significant factors in a particular nation.

The third major option for placement of forestry responsibility is in a separate ministry, often called "Natural Resources" or "Forestry and Natural Resources." In some situations this includes minerals and mining, and this can have detrimental effects on the priority given to forestry. In most cases, however, this separate setting strengthens the position of the forestry focus.

Such ministries, however, seldom compare well with agriculture in the competition for funding, facilities, and people. The dominance of forestry professionals within the ministry does have a positive effect on esprit de corps. Foresters are at least not second-class citizens in their own organization. Their own professional norms occupy a more prominent position in the organizational mythology.

This may also lead to conflict with other organizations. For instance, protecting local hardwood forests from depletion may require confrontation with foreign exchange–earning extractive industries (such as mining) or wood energy–intensive agricultural activities (such as tobacco production). Or it may pit foresters against national elites bent on transforming local natural wealth into foreign financial deposits.

This organizational form allows conservation perspectives to attain an equal status with production perspectives. But poor links to village organizations and an underdeveloped extension system often characterize this model. (In one country a local forester called the forestry extension system a "tree without roots.")

The fourth major location for the public sector unit charged with responsibility for forestry management is in a ministry of "Natural Resources and Tourism." This tends to link trees and wealth in a very different way than do the other options.

When countries have unique, special, or abundant fauna that attract global attention, there is often a tourist industry based on that resource. Tourism generates foreign exchange through the preservation of the resource rather than its extraction (although the ivory traffic and the trade in endangered species do represent illicit, short-term extractive behavior). Thus there is at least some pressure toward afforestation, species conservation, and the treatment of indigenous species of flora and fauna as national resources.

This introduces an economic value for forests and wildlands as species habitats, an emphasis less common among the other organizational options. Appreciation for indigenous tree species and an ecological perspective are more respectable in such surroundings than in some of the others described above.

The tourism trade can also be a trap. It tends to be import intensive, it requires a strong world economy to guarantee international visitors, and it is usually an enclave industry benefiting mainly a few urban entrepreneurs. Without institutional mechanisms to share revenues with villagers in areas adjacent to wildlife reserves or development investments to limit the need for encroachment, the link to tourism can be risky. Other concerns such as soil erosion, moisture retention, and forest protection in areas removed from tourist views may be slighted.

Likewise, system-wide procedures and behaviors need to be understood to avoid performance problems. The internal design of forestry management organizations is only one part of the institutional puzzle. The balance of institutional agenda, resources, and linkages is also required.

THE BALANCE OF INSTITUTIONAL FORCES

The operation of the formal forestry management organizations noted above will vary with two sets of factors. One is the overall set of public institutions in relation to the forestry organization. The other set involves cultural patterns and the informal distribution of power and incentives within the nation.

Interagency Dynamics

The balance of resources and agenda among ministries will shape the operation of a forestry management organization. For example, in Thailand, where the forests are so depleted that logging is now illegal, the Ministry of Forests is charged with protection of forests and national parks. At the same time, there is a conflict with a ministry that would seem to be a natural ally—the Ministry of Tourism.

Rather than promoting ecotourism, the Ministry of Tourism sees its mission as championing the construction of roads and artificial lakes to attract the Asian tourist trade and capture foreign currency. This may be an accurate reflection of present demand for leisure opportunities among the wealthy classes of Asia, but it directly pits one ministerial agenda against another. Relative budgetary strength and access to sympathetic and powerful actors, such as the king, will be important determinants of which perspective dominates.

In other countries the interorganizational dimension takes a similar but not identical form. Relative budgets reveal much about true priorities. When a forestry department devotes most of its budget to a production division and that division of one department has a budget that greatly overshadows the total budget of a regulatory department, such as Papua New Guinea's Department of Conservation and Environment, then the task of the second organization is made much more difficult. This particular scenario is played out in many tropical countries, as well as in some major nations of the northern hemisphere.

Overlapping jurisdictions among the various ministries charged with forest protection and utilization also can complicate the picture. In some countries mangrove and other coastal forests come under the authority of a different ministry than do upland forests. Or contradictions among the roles of local governments and departments of agriculture, forestry, and public works can delay the identification of authority to stop questionable practices until it is too late and the forest is a thing of the past.

Unclear division of responsibility and authority serves the interests of exploiters in cases where the strongest leadership resides in the organizations sympathetic to the mining of the wood. Where the stronger leader heads an organization aiming at protection and sustainable use, then the protection agenda might temporarily prevail as a result of fuzzy jurisdictions. Protective action can precede sorting out of the legalities, but this is not often the case. Generally, clarification of responsibility will help to save the forests. Nevertheless, each case must be assessed based on local circumstances.

Other system-wide practices such as reimbursement procedures for civil servant expenses or monthly payment systems for all ministries can greatly constrain the activities of forest departments but not parastatal bodies. Sometimes important factors come from across national boundaries. For example, Thailand has banned logging but not the transport of logs. So timber from Laos and Burma makes its way through Thailand—deforestation has been exported.

Formal relationships among public and private organizations are not the only complicating factors. Often informal political agreements and social practices change the nature of organizational dynamics.

Informal Dynamics

The formal allocation of power and authority does not always reflect true decision-making practices. For instance, if the military acts as a shadow government and has veto power over development activity, then the military must be educated and co-opted into helping to save the forests. Ignoring the military could negate the effort in terms of both poorly mobilized efforts and contradictory policies, such as the construction of security roads into virgin forest and protected areas.

Informal behavioral systems also can be important. Well-intended attempts at regulation may produce only unintended opportunities for levying bribes on those to be regulated, or the formal mission attributed to an organization may be at odds with its actual operating style. Attempts to rein in deforestation dynamics will be more likely to succeed if such factors are taken into account.

The culture of an organization may be at odds with its formal mission. Viewing public employment as a chance to eat public resources is one version of this that is found throughout the world. Structural factors can reinforce nonprofessional behavior. For example, the need to move from a technical to a managerial career track once a certain level has been reached can dilute the limited expertise within an organization. Experimentation with uncapped technical promotion paths is being conducted in Asia under the rubric of "functional organization."

The Bureau of Forest Development (BFD) in the Philippines shows the importance of informal factors. Although its formal mission was to protect the forests and regulate logging, its informal operation was as an assessor of royalties and protector of logging interests. Under Marcos its staff lived in comfort based on the bribes they extracted from illegal loggers. In attempting to clean up this situation, the present administration has recognized the need to dismantle the BFD clique. Senior people have been retired, some junior people have been dismissed, and most of those who remain in the civil service have been transferred to other organizations. In addition, a new department has become the home of the previous BFD functions. Thus, both formal reorganization and informal personnel dispersal are recognized as necessary ingredients for reorienting a bureaucratic entity.

The discussion above suggests that the role of institutional factors in tropical deforestation is complicated and sometimes counterintuitive. Nevertheless, it is real and major. One observer of Third World forestry initiatives stated that "critique after critique within and among countries and bilateral and multilateral donors point to institutional failures as major obstacles to success. . . . So serious are the questions of institutional development that, if donors and country governments are unwilling to confront them realistically, then they face the harsh dilemma of postponing the programs until they are" (Buckman 1987, 121). Clearly, any donor strategy to slow down deforestation must contain a core component emphasizing institutional strengthening, reform, or innovation. To work, the strategy must take cognizance of both formal and informal dimensions of organizational behavior. But some of the strengthening, reform, and innovation may need to occur within the donors themselves.

Donor Dynamics

International public and donor pressure is having an effect. Indonesia, for example, is very sensitive to a perceived world image of it as a country squandering its resources. Continuing low oil prices, however, intensify the search for foreign exchange–earning opportunities, and forest depletion continues apace.

Well-intentioned donor programs can even hasten the depletion process. For example, installing plantations on degraded lands gives an incentive to mine it now, declare it degraded, and then get funding to establish an estate of exotic species. Investment in rehabilitation can thus encourage immediate destruction.

Often the donor message is even more mixed than this. In some countries both bilateral and multilateral aid programs are increasing their forestry focus. But this is not a new infusion of resources—it is merely a reordering of priorities within a previously set program level. (An exception is the Japanese fund that is presently being established, but some observers see this as little more than "guilt money" to compensate for the role of the Japanese private sector in tropical deforestation.)

Sometimes the new donor emphasis is only cosmetic in symbolic ways. Changing a project name to include *forestry* or *natural resources* does not necessarily reflect a reordering of priorities and resources. Likewise, ignoring needs such as municipal water systems or urban housing may be passing up an opportunity to lessen some pressure on the forest land, even though these projects contain no mention of forests. Substantive connections among programs, not titles of loans or grants, are key to reducing the rate of tropical deforestation. A coherent and comprehensive strategy is needed. Even the Tropical Forestry Action Plan has been criticized for overemphasizing forest industry and slighting other dimensions.

There are clearly many actors and many levels of involvement where improvements are needed. There is no institutional silver bullet to solve the problem of tropical deforestation, nor is there any single ogre responsible for the problem. Without better institutional mapping and an increased focus on institutional change, natural resource management performance is not likely to improve. Both institutional studies and institutionally oriented technical assistance are needed to strengthen the entire range of institutions involved in sustaining or protecting the planetary tropical forest resource.

STRENGTHENING INSTITUTIONS TO SAVE
TROPICAL FORESTS

The foregoing assessment suggests two things. First, both direct and indirect causes of deforestation must be addressed for a long run solution. Second, solutions will require selective strengthening or reorientation of existing institutional configurations and operations, or maybe the invention of new institutions, to work. All three levels—international, national, and local—will be affected.

Dealing with this complex problem will entail embarking upon two different solution paths simultaneously. The first path leads toward limiting demand for wood consumption and forest clearing. This is a long run emphasis on technology development and diffusion and on public awareness. The second path leads toward protecting the supply. Short- and medium-term sustainable uses of the forest resource and direct protection of the forest from human predation characterize this path. All of these contain institutional aspects, and those aspects are noted below.

LIMITING DEMAND

A basic element of demand is population growth in the consuming countries. Ultimately, this must be checked. In the short run, however, there are three key areas to be addressed. One is the development and diffusion of energy-saving and substitute technologies to limit demand for wood fuel and construction material. Another involves environmental education to shape values that are more conducive to less exploitative uses of forest resources. The third is direct suppression of trade in forest products.

Technology Development and Diffusion

Technologies are needed to increase the efficiency of wood use, provide substitutes for wood products, improve harvesting methods to alleviate the destruction of forests when timber is extracted, promote agroforestry, allow the production of indigenous species, and manage human population levels. Indeed, the list could go on. The point, however, is that institutions are needed to conduct research, develop prototype equipment and practices, test them, and diffuse them to users. Most of the institutions presently existing in the Third World have limited capacities to do this.

One part of an agenda for institutional strengthening would be an emphasis on technology development organizations. Better staff, equipment, supplies, and facilities are needed, but reorientation of effort and focus may also be necessary. Knowing and respecting village viewpoints may spell the difference between success and failure. Technology development may be done well in a public sector setting, but manufacturing and marketing may be better left to the private sector. Development, however, would need linkages with producers and marketers to understand their constraints and allow adjustment of technical designs.

This suggests that not just a few organizations but rather technology development institutional networks are needed. To work, technology development and diffusion requires multiple actors and organizations. Subsidies of research and development work and policies to encourage the development and use of new technologies may also be needed.

Education and Values

Conservation education is considered to be an integral part of any conservation program. Appreciation and respect for nature is depicted as a *sine qua non* for generating public support for conservation programs and instilling an appropriate behavioral ethic in a population. For the long-term perspective this is undoubtedly correct.

But let us not be naive. Unless individuals, communities, and organizations are provided beneficial alternatives to forest-destroying practices, the destruction will continue because their livelihoods depend upon it. People do not destroy natural resources just because they want to, but rather because they see themselves as having to. Without opportunities to earn a living in nonharmful or restorative ways, people will continue to damage forests even if they are aware of the dangers in doing so. Mass campaigns of education and rhetoric will have little effect unless there are opportunities and rewards for different behaviors.

For those whose earnings are not directly dependent on forest exploitation, however, education may be a wedge into bureaucratic inertia. Information sessions with policy makers may help to persuade them of the need for new policy and program efforts. They may be able to help restructure the national reward system to encourage positive action. Knowledge of the concepts of natural resource accounting, visible evidence of watershed deterioration, firsthand exposure to the effects of landslides from deforested slopes, and pressure from donors can all be brought to bear on policy developers.

One element of education and values is family planning. Not only must limitation of human population growth be vigorously pursued, but also the lessons learned over the last twenty years in social marketing and diffusion of family planning practices and technologies should be adapted for tropical forest protection. Successful family planning programs, such as Indonesia's, may be a gold mine of strategic and tactical information.

Market Suppression

Another approach to limiting demand for forest destruction is to suppress the market forces. Recent experience with the ivory trade provides an encouraging example of an effort to do this, although experience with limiting the harvesting of whales offers a discouraging one. Helping nongovernmental organizations (NGOs), supporting policies that bar market access, and policing international trade in forest products can be elements of market suppression.

Convincing international financial institutions to withhold credit or insurance coverage for ventures that threaten the forests would be another aspect of market suppression. Anything that increases, directly or indirectly, the costs of destructive behavior should be explored.

PROTECTING SUPPLY

Two basic thrusts will be needed to protect the dwindling supply of tropical forest cover. The first emphasizes sustainable utilization. The second emphasizes direct protection and nonutilization of the forest products.

Social Forestry and Sustainable Use

There is increasing evidence from the Amazon to Australia that, over the long run, tropical forests yield high financial returns when they are farmed for their nonwood products instead of mined for their timber. Likewise, selective and careful extraction of forest wood that maintains the integrity of the forest ecosystem can yield greater long-term profits than monoculture plantation production that destroys the ecosystem. But this is the long term. The problem is that wood mining generates fast profits.

A key to solving this problem is getting control of the forest resource into the hands of those who have a vested interest in maintaining a sustainable yield system and out of the reach of those whose interests are in the immediate exploitation of the resource. This is no easy task. It will require strong policy shifts, organizations capable of enforcing the policies and protecting them against political and economic market pressures, and the development of grassroots institutions to manage the forest resource.

An approach increasingly popular among international donors is to build grassroots organizations to manage sustainable forest enterprises, grow trees for household fuel and construction needs, conduct research on exotic and indigenous species, and examine policy prerequisites for sustainable forestry management. When these elements are placed under a single program umbrella, the result is often called *social forestry*. Such programs are increasingly confronting performance problems of an organizational nature.

Social forestry is not simply the growing of trees by forestry technicians. Quite the opposite—*social* comes before *forestry*. That is, the effects on the trees and other natural resources are achieved only through the actions of nonforesters. Conserving woodlands, sustainably harvesting forest products, planting and nurturing woodlots, and improving the efficiency of wood-consuming activities all result from the actions of people outside forestry agencies—villagers, entrepreneurs, policy makers, local leaders, court officers, and researchers are among the key actors. Indeed, a social forestry program achieves its objectives by working with and through various individuals, social groups, and formal organizations rather than by controlling the performance of foresters.

At the same time, the training of foresters emphasizes technical skills and plantation management methods—not extension, negotiation, or leadership. Moreover, forestry departments are seldom structured to emphasize operations that link them to external organizations. But with social forestry this is key.

The social part of social forestry involves working closely with villagers as well as other organizations. Rural knowledge, attitudes, and practices may be central to success. For instance, in one country farmers resisted planting *gmelina* when they saw that it resulted in a "dead forest" without birds or other wildlife. In another country, perspectives on tree production were gender related, with women viewing trees in terms of firewood potential and men seeing them as prospective poles for houses and fences.

Institutional strengthening in the area of social forestry will require the development of new research capabilities and resources, it will contain analytical and legislative activities concerning land and tree tenure systems, it will support grassroots organizational development, it will encompass reorientation of pricing policies and marketing infrastructure and policies, it will involve the development of incentive systems that promote a performance-oriented organizational culture as opposed to an extractive culture among key institutions, it will entail training foresters and protected area staff in new skills and approaches, and it will include reorganizing implementing agencies and providing technical assistance to help them to leverage the resources of other organizations. Creating linkages between training institutes, local communities, and policy makers will also be part of this.

There have been recent innovations in methods for creating these linkages. For example, action planning workshop approaches using stakeholder analyses and coordination/negotiation exercises with real workgroups have registered success in numerous countries. Such methods will be integral to activities focusing on social forestry and sustainable yield management of forest resources.

Protected Areas and Buffer Zones

Although advances are being made in the area of reclamation science, the cost of protecting an existing forest is much less than the cost of reconstructing a destroyed one. Thus, for the immediate future, approaches that maintain selected existing forest ecosystems are likely to be far more cost effective than those employing reconstructive tactics.

The immediacy and magnitude of the threat to tropical forests dictate that one component of a plan for minimizing that threat must be direct protection—the establishment of parks and reserves that constrain, and sometimes prohibit altogether, extractive activity. This also has institutional dimensions.

Some of the capacity-building requirements are straightforward and obvious—action-oriented training, increased staffing, and the provision of equipment and supplies to organizations entrusted with the management of protected areas are among them. Likewise, strengthening the training and outreach roles of the institutes educating foresters, park guards, and natural resource managers is equally straightforward.

When debt-for-nature swaps are negotiated, the design and development of the organizations overseeing and implementing the reserves will also be important—organizational "trees without roots" can turn short-term solutions into long-term disasters. Enforcing compliance with the terms of the debt swap will require strong institutions able to marshall allies and support from multiple segments and levels of society and able to keep that alliance over the long run.

When a local NGO is given a monopoly over the management of a protected area, care must be taken to match the scale of the management task to the scale of the local organization. The NGO will also need a secure funding source for the future and allies with aggregate power exceeding that of likely coalitions of predator organizations. Otherwise, defeat will only have been delayed, not diverted.

Other capacity-building efforts, however, are directed at organizations without a direct management role and are less obvious. For example, the strengthening of court systems and prison facilities may be needed along with new laws reflecting the newfound appreciation for preservation of species and habitats. Similarly, policy analysis capability to consider such things as the demographic influence of job creation in urban settings away from protected areas, the effect of roadbuilding in adjacent provinces, land tax preferences for developing new land, and the effect of price policies on tourism may be less obvious but equally important. And the organizations with power may be outside the normal constellation of development and conservation institutions. In some countries the military is a crucial actor and the operational definition of national security may exclude forest preservation.

Much of the institutional focus will be on buffer zone and transition zone management. This is an area that is both new and lacking in knowledge and, at the same time, old and commanding a wealth of knowledge. The new part involves walking a tightrope—ensuring adequate economic opportunity in the area surrounding the reserve to alleviate the need for people to encroach on the reserve to obtain a livelihood, while simultaneously not creating a growth pole or magnet that draws people into the area. How to do this is not presently known. The old part involves much of the accepted wisdom on rural development management. The lessons learned over the past two decades about participation, decentralization, local revenue administration, integrated rural development and administrative control, local organizations, land tenure, gender considerations, small-scale enterprise, social learning, and the processes of rural development management all hold some of the keys to effective transition zone management. It is not necessary to reinvent the wheel.

Indeed, some of the existing knowledge may be found in places far afield from protected areas. Improving the capabilities of higher and higher density urban areas to keep people away from the remote zones may be crucial for protecting those zones. Rural protection may be a direct function of success in urban development. This may be especially true in those cases where the transition from urban area to buffer zone is a quick one.

Linked to protected area management is the tourism trade as a revenue generator. From East Africa to the Galapagos, ecotourism has helped to protect and give economic value to critical ecosystems, but most tourism has down sides as well. First, it must be managed so as not to destroy the value of the area. A stream of tourists can cause as much erosion as a stream of water. Second, international tourism is fickle, requiring a strong world economy to generate traffic. Third, tourism is not only a source of foreign exchange, it also consumes large amounts of foreign exchange itself. And fourth, tourism is an enclave industry with a high percentage of its proceeds going to an urban elite. Experimentation with revenue sharing with localities on the fringe of the protected area is occurring in some countries (such as Zimbabwe), but much more needs to be done to ameliorate the negative effects of the tourist trade. Building local, or regional, tourism is a partial answer, but it is usually inadequate and it produces less foreign exchange.

Creating incentives for nondestructive behavior and mobilizing popular support for sustainable development are keys to success. Policy reform related to these thrusts is also important, and strengthening policy analysis capabilities is concomitant with all three emphases. Natural resource economists and conservation biologists certainly need access to national policy makers. But implementation capacity must follow policy reform. Otherwise, rhetoric reigns and the forests continue to disappear.

CONCLUSION: THE INSTITUTIONAL IMPERATIVE

Putting the brakes on tropical deforestation requires three institutional initiatives—mapping, strengthening, and inventing. These are discussed as they relate to the issue of sustainable forestry in North America.

INSTITUTIONAL MAPPING

Three types of institutional studies are needed. First, comparative data need to be developed to allow assessments of relative institutional strengths and weaknesses among sectors and countries. For instance, the percentage of the public budget or public wage bill devoted to natural resource enhancement and protection could show comparative commitments to environmental improvement, but other, complementary data might be needed to interpret this fully.

For example, the percentage of a resource management agency's budget committed to personnel costs, or the wage bill, would give an idea about absorptive capacity and ability to respond creatively to new challenges. In the United States, state governments typically spend about 70 percent of their recurrent budgets on staff cost. In the Third World, however, the corresponding number is closer to 95 percent. This has clear implications.

Among sectors, some ratios might also be useful. For example, the ratio of foresters to agricultural extension agents or teachers or soldiers might reveal relative priorities in the past, imbalances with future needs, and possible human and organizational resources to be tapped in the effort to halt deforestation.

Comparisons of the openness of information flows in different countries could also be useful. Control of the press and the accompanying secrecy of exploitation make a difference in the ability of donors, NGOs, and political opponents to mount a campaign against resource degradation practices.

Comparative data on tax policies that influence resource use would also be helpful. The presence or absence of key policies and incentives and the relative strength of important government agencies and NGOs need to be mapped far better than they have been to date. Simple descriptions and identifications are needed before more sophisticated analyses can be done.

The second type of study that should be conducted is a case study. Identifying how the institutional configuration evolved and what informal mechanisms affect performance can better be done through case studies by institutional specialists with a grounding in the natural resource management field.

Care must also be exercised here. A core set of common data points covering key institutional relationships and human behaviors must be part of the case study exercise. Although it remains largely uncodified, there is already enough field experience among institutional specialists to suggest what that core data set should include.

The applied purpose of the exercise should also be kept in the forefront. Extracting possible lessons for replication elsewhere and aggregating cases to discover influence patterns would both complement comparative data development and further the art of technical assistance to natural resource management institutions.

The third type of institutional study involves examining traditional resource management systems used by indigenous people throughout the planet. Understanding the connections among natural resource bases, social values, political institutions, and sustainable agriculture can be invaluable. Tapping a "cultural gene pool" may uncover new strategies for sustained-yield and multiple-use forestry management.

Key mapping thrusts should be

- comparing the openness of information and relating it to the success of NGOs, the performance of natural resource agencies, the effectiveness of regulation, and so forth, and
- developing a comparative worldwide data base of forestry resource management institutions using indicators such as budgets, staff, percentage of budget devoted to wage bill, jurisdictional overlaps with other agencies, national policies, performance incentives, analytical units, and so forth.

INSTITUTIONAL STRENGTHENING

Some of the case studies could be conducted as part of technical assistance exercises or donor sector assessments. When combined with action planning efforts, these studies could directly help natural resource management agencies to perform better.

For example, a public sector management review could focus on the natural resource management subsector. Institutional barriers to following soft energy paths and reclaiming degraded environments could be integral to such an effort, and an understanding of institutional constraints, opportunities, and linkages at the national, intermediate, and community levels could help to promote more informed and effective assistance in natural resource management.

Integrating institutional specialists into technical teams and providing greater management assistance to resource management programs could also combine performance improvement with knowledge generation. Techniques are available to help resource technicians deal with institutional problems transcending their own organizations and sectors. Such help should be increased dramatically. But this is not enough. A coherent, concerted effort to develop a data base on natural resource management institutions is long overdue. Forestry could become the pilot sector for developing such a data base.

Key thrusts in operational assistance should be

- changing the relative resource strength of natural resource training, protection, and monitoring agencies in relation to exploitation agencies;
- strengthening performance-oriented incentives within protection agencies;
- assisting with reorganizing the balance of public sector resource management agencies and clarifying jurisdictions of those agencies;
- generating opportunities for villagers to obtain livelihoods in nondestructive ways;
- reviewing land tenure systems and arrangements and their effects on deforestation and promoting forest-conserving land use and land distribution;
- developing natural resource policy analysis units in ministries of finance; and
- building capacities within protection agencies, subnational government and community organizations, and NGOs to manage their resources and influence other actors.

INSTITUTIONAL INVENTION

The discussion above emphasized the importance and nature of institutional elements in the use and misuse of tropical forests. A constant theme has been the need to understand more fully the roles played by institutions to date and to incorporate this knowledge into any strategies developed to combat deforestation. Indeed, it is argued that institutions must play key roles in such strategies.

But let us not narrow our vision by considering the institutional arrangements of the past as the full array of options for the future. We do not need just to do better what we have done—we also need to do new things. And we need new institutional arrangements to do new things.

Numerous institutional inventions have affected human history—the nation-state, the bureaucracy, the private corporation, the political party, the election of leaders, the research university, the bank, the agricultural extension system, and the cooperative society, among others. They emerged as responses to the need for new ways of ordering human relationships to achieve objectives or execute specific tasks. At one point they were innovations. Once they became familiar aspects of the social landscape, it became hard to imagine a world without them or to see alternatives to them.

In times of crisis imagination is called for, and this may be such a time. The present array of institutions seems hard pressed to respond to the planetary environmental degradation we are witnessing. We use organizational mechanisms not because they are necessarily what we need but because they are there. We choose when we should invent.

The present emphasis on NGOs may be a case in point. The search for alternatives to bureaucratic stasis once pointed to parastatal bodies as a potential alternative. Now they are in disrepute, and the NGO has risen as the new hope.

Innovations such as the debt swap are needed. Public trusts, legal standing for trees, and various other basic changes are promoted by different actors. Some may prove to have staying power, whereas others may be little more than short-lived experiments. But such experiments are necessary if we are to meet the needs of the end of the twentieth century.

The needs that we have identified include the inhibition of human reproduction, the empowerment of local communities, the provision of sound scientific knowledge to local organizations, the buffering of poor people from the pressures of international marketeers, the generation of economies based on sustainable processes and principles, the equitable redistribution of access to land, and the redefinition of security to include a supply well into the future of the benefits generated naturally from the tropical forests. Both public bureaucracies and NGOs certainly have a role to play in meeting these needs, but we may have to cast away blinders that limit our views—blinders like the public/private dichotomy, national sovereignty, the mutually exclusive definitions of conservation and development, or the segregation of the natural and social sciences.

Additionally, we need to go beyond the currently popular views of the evils of subsidies or regulation. Indeed, a regulatory rebirth based on the offering of carrots instead of the brandishing of sticks may be necessary. Just as coproduction of services by users and providers has helped to redefine organizational roles in many areas, so too the sharing of regulatory, protective, and distributive functions may be the key to sustainable natural resource management in the twenty-first century.

This synopsis of experience from the tropics suggests seven areas of concern for the promotion of sustainable forestry in North America.

1. Population policy: Unless population growth is checked, the pressures for overexploitation will increase to the point where they are uncontrollable.
2. Open information flow: Media and public knowledge of what is happening is essential to keep actions on the right track.
3. Tax codes and public policies: Human behavior and land use are often determined by incentives created by factors not directly focusing on natural resources, and these must be made compatible with sustainable resource use requirements.
4. Private policies: Employment practices of private companies can lead to stress on the natural resource base. Lack of fringe benefits for part-time employees or unacceptance of telecommuting, for example, can reinforce resource squandering. Private practices should be examined for their resource implications.
5. Reorientation of private organizations: Some organizations that earn their profits exploiting the resource may be able to be transformed into organizations that keep some of their skills and attributes but apply them to nonexploitative activities, and assistance should be given to explore such transformation possibilities.
6. Reorientation of public organizations: Public natural resource management agencies often take narrow technical perspectives toward their missions. They need to become broader in terms of embracing new skills and coproducing services with a wide range of other actors, and they need assistance in making the transition.
7. Creation of ecomarkets through public investment: Using public funds and policies to encourage research and development, goods and service production and delivery, and land use that support a sustainable society is an essential element in institutional reorientation.

These seven institutional recommendations confront both barriers and possibilities. They go beyond the knowledge hypothesis and support new behavior. The challenge is to transform them into institutionalized actions.

SELECTED BIBLIOGRAPHY

Agarwal, B. 1986. *Cold hearths and barren slopes: The woodfuel crisis in the Third World.* London: Zed Books.

Ahmad, Y., S. El Serafy, and E. Lutz. 1989. *Environmental accounting for sustainable development.* Washington, DC: World Bank.

Anderson, R., and W. Huber. 1988. *The hour of the fox: Tropical forests, the World Bank and indigenous people in central India.* Seattle: University of Washington Press.

Arnold, J. E. M. 1987. Community forestry. *AMBIO—a Journal of the Human Environment* 16(2-3):122-128.

Blaikie, P., and H. Brookfield. 1987. *Land degradation and society.* New York: Methuen.

Bromley, D., and M. Cernea. 1989. *The management of common property natural resources: Some conceptual and operational fallacies.* Discussion paper 57. Washington, DC: The World Bank.

Buckman, R. E. 1987. Strengthening forestry institutions in the developing world. *AMBIO—a Journal of the Human Environment* 16(2-3):120-121.

Cernea, M. 1985. Alternative units of social organization sustaining afforestation strategies. In *Putting people first: Sociological variables in rural development,* ed. M. Cernea, 267-293. New York: Oxford University Press.

Chambers, R., and M. Leach. 1989. Trees as savings and security for the rural poor. *World Development* 17(3):329-342.

Chambers, R., N. C. Saxena, and T. Shah. 1989. *To the hands of the poor: Water and trees.* Boulder, CO: Westview Press.

Clarke, J. N., and D. McCool. 1985. *Staking out the terrain: Power differentials among natural resource management agencies.* Albany: State University of New York Press.

Clay, J. 1988. *Indigenous peoples and tropical forests: Models of land use and management from Latin America.* Cambridge, MA: Institute for Cultural Survival.

Denslow, J. S., and C. Padoch, eds. 1988. *People of the tropical rain forest.* Berkeley: University of California Press/Smithsonian Institution.

Dewees, P. 1989. The woodfuel crisis reconsidered: Observations on the dynamics of abundance and scarcity. *World Development* 17(8):1159-1172.

Du Toit, R. F., B. M. Campbell, R. A. Haney, and D. Dore. 1984. *Wood usage and tree planting in Zimbabwe's communal lands: A baseline survey of knowledge, attitudes and practices. A report to the Forestry Commission of Zimbabwe and the World Bank.* Harare: Resource Studies.

Edington, J., and M. A. Edington. 1986. *Ecology, recreation and tourism.* Cambridge: Cambridge University Press.

Food and Agriculture Organization of the United Nations, World Bank, World Resources Institute, and United Nations Development Programme. n.d. *The tropical forestry action plan.* Rome: FAO.

Goulden, R. 1989. *Thoughts on change for resource managers: Transactions of the Fifty-fourth North American Wildlife and Natural Resources Conference,* 611-615. Washington, DC: Wildlife Management Institute.

Gradwohl, J., and R. Greenberg. 1988. *Saving the tropical forests.* London: Earthscan.

Grainger, A. 1988. Tropical rainforests—global resource or national responsibility? In *For the conservation of earth,* ed. V. Martin, 94-99. Golden, CO: Fulcrum.

Gregersen, H. 1990. *Key forestry issues facing developing countries: A focus on policy and socioeconomics research needs and opportunities.* St. Paul: University of Minnesota Forestry for Sustainable Development Program.

Gregersen, H., S. Draper, and D. Elz. 1989. *People and trees: The role of social forestry in sustainable development.* Washington, DC: World Bank.

Harrison, P. 1987. *The greening of Africa: Breaking through in the battle for land and food.* New York: Penguin Books.

Heath, R. 1986. The National Survey of Outdoor Recreation in Zimbabwe. *Zambezia* 13(1):25-42.

Heberlein, T. 1988. Improving interdisciplinary research: Integrating the social and natural sciences. *Society and Natural Resources* 1(1):5-16.

Hirschman, A. O. 1970. *Exit, voice and loyalty: Responses to decline in firms, organizations and states.* Cambridge: Harvard University Press.

Honadle, G. 1982. Rapid reconnaissance for development administration: Mapping and moulding organizational landscapes. *World Development* 10(8):633-649.

———. 1989. *Putting the brakes on tropical deforestation: Some institutional considerations.* Washington, DC: Agency for International Development.

———. 1989. Interorganizational cooperation for natural resource management: New approaches to a key problem area. In *Transactions of the 54th North American Wildlife and Natural Resources Conference,* 271-276. Washington, DC: Wildlife Management Institute.

Honadle, G., and J. Vansant. 1985. *Implementation for sustainability: Lessons from integrated rural development.* West Hartford, CT: Kumarian Press.

Hughes, F. 1987. Conflicting uses for forest resources in the lower Tana River basin of Kenya. In *Conservation in Africa: People, policies and practices,* eds. D. Anderson and R. Grove, 211-228. Cambridge: Cambridge University Press.

Hurst, P. 1990. *Rainforest politics: Ecological destruction in South-east Asia.* London: Zed Books.

James, J., and E. Gutkind. 1985. Attitude change revisited: Cognitive dissonance theory and development policy. *World Development* 13(10/11):1139-1149.

Jordan, C. 1986. Local effects of tropical deforestation. In *Conservation biology: The science of scarcity and diversity*, ed. M. Soulé, 410-426. Sunderland, MA: Sinauer Associates.

Klee, G., ed. 1980. *World systems of traditional resource management*. New York: Halsted Press.

Korten, D. 1987. Third generation NGO strategies: A key to people-centered development. *World Development* 15(suppl):145-160.

Ledec, G. 1985. The political economy of tropical deforestation. In *Divesting nature's capital*, ed. H. J. Leonard, 179-226. New York: Holmes & Meier.

Ledec, G., and R. Goodland. 1988. *Wildlands: Their protection and management in economic development*. Washington, DC: World Bank.

Leonard, H. J., M. Yudelman, J. D. Stryker, J. O. Browder, A. J. DeBoer, T. Campbell, and A. Jolly. 1989. *Environment and the poor: Development strategies for a common agenda*. Washington, DC: Overseas Development Council.

Little, P., and D. Brokenshaw. 1987. Local institutions, tenure and resource management in East Africa. In *Conservation in Africa: People, policies and practice*, eds. D. Anderson and R. Grove, 193-210. Cambridge: Cambridge University Press.

Mahar, D. 1989. *Government policies and deforestation in Brazil's Amazon region*. Washington, DC: The World Bank.

McGranahan, G. 1991. Fuelwood, subsistence foraging, and the decline of common property. *World Development* 19(10):1275-1287.

Moser, M. 1989. Recent successes in international wetland conservation. In *Transactions of the Fifty-fourth North American Wildlife and Natural Resources Conference*, 75-80. Washington, DC: Wildlife Management Institute.

Noronha, R., and J. Spears. 1985. Sociological variables in forestry project design. In *Putting people first: Sociological variables in rural development*, ed. M. Cernea, 227-266. New York: Oxford University Press.

Office of Technology Assessment, U.S. Congress. 1984. *Technologies to sustain tropical forests*. Washington, DC: US Government Printing Office.

Ostrom, E. 1987. Institutional arrangements for resolving the commons dilemma: Some contending approaches. In *The question of the commons: The culture and ecology of communal resources*, eds. B. M. McKay and J. M. Anderson, 250-265. Tucson: University of Arizona Press.

Peters, W., and L. Neuenschwander. 1988. *Slash and burn: Farming in the Third World forest*. Moscow: University of Idaho Press.

Poffenberger, M. 1990. *Keepers of the forest: Land management alternatives in Southeast Asia*. West Hartford, CT: Kumarian Press.

Repetto, R. 1987. Creating incentives for sustainable forest development. *AMBIO—a Journal of the Human Environment* 16(2-3):94-99.

Repetto, R., and M. Gillis, eds. 1988. *Public policies and the misuse of forest resources*. Cambridge: Cambridge University Press.

Romm, J. 1986. Frameworks for governmental choice. In *Community management: Asian experience and perspectives*, ed. David Korten, 225-237. West Hartford, CT: Kumarian Press.

Silverman, J. M. 1990. *Public sector decentralization: Economic policy reform and sector investment programs*. Division study paper 1, Public Sector Management Division, Africa Technical Department. Washington, DC: World Bank.

Southgate, D., R. Sierra, and L. Brown. 1991. The causes of tropical deforestation in Ecuador: A statistical analysis. *World Development* 19(9):1145-1151.

Spears, J. 1988. Preserving biological diversity in the tropical forests of the Asian region. In *Biodiversity*, ed. E. O. Wilson, 393-402. Washington, DC: National Academy Press.

Tendler, J. 1989. What ever happened to poverty alleviation? *World Development* 17(7):1033-1044.

Thomson, J. T. 1988. Deforestation and desertification in twentieth-century arid Sahelian Africa. In *World deforestation in the twentieth century*, eds. J. Richards and R. Tucker, 70-90. Durham: Duke University Press.

United Nations Economic and Social Commission for Asia and the Pacific. 1986. *Environmental and socio-economic aspects of tropical deforestation in Asia and the Pacific*. Bangkok: ESCAP.

World Resources Institute and International Institute for Environment and Development. 1986. *World Resources 1986*. New York: Basic Books (and subsequent editions: 1987, 1988-89, 1990-91).

PART II

Designing Sustainable Ecological
Systems: A Regional Approach

Introduction

Gregory H. Aplet and Jeffrey T. Olson
The Bolle Center for Forest Ecosystem Management,
The Wilderness Society

One of the central issues driving the search for sustainable forestry both domestically and abroad is the unfortunate fact that the ecological systems upon which forestry depends have limits. By neglecting certain aspects of these systems in the pursuit of commodity extraction, we have exceeded those limits with respect to many aspects of the forest ecosystem. We now know that we must move beyond the simple management of commodities, even defined through the notion of multiple use, to the management of ecological systems.

Management of ecosystems, however, represents a seemingly overwhelming challenge. Ecological systems occur at all spatial scales from the microsite to the landscape or region and transcend human time scales. They encompass all species, many of which are yet to be discovered, and they encompass the countless interactions among those species and interactions with the physical environment. It is no wonder, then, that the transition from commodity management—the focus on a few commercially valuable species—to ecosystem management has been slow to develop.

Despite these impediments, tremendous progress has been made in the last decade or so toward theoretical and practical approaches to ecosystem management. Perhaps more than anywhere else, the crucible of this change has been the Pacific Northwest. Dwindling ancient forests, endangered species issues, and an economy in transition have forced managers and researchers to seek innovative solutions to forest management problems. The challenges of the Northwest have inspired Jerry Franklin to become one of the true innovators of ecosystem management. In his introduction to the fundamentals of this new approach, Franklin lays the foundation for a "kinder, gentler forestry." The objective of this New Forestry is to achieve long-term sustainability, which Franklin defines as the maintenance of the *potential* to attain any desired state. This requires maintaining all of the parts (genetic potential) and all of the processes (productivity).

While discussing his five fundamentals of ecosystem management, Franklin builds an incontrovertible case for the role of the matrix, the managed landscape that forms the "sea" between the "islands" of reserved lands. He argues that reserves can never effectively conserve all aspects of biodiversity because they can never cover all relevant scales. Different organisms function at different spatial scales and respond to signals that transcend reserve boundaries. Sustainable management of the landscape requires that the same aspects of structure and function that occur in reserved lands also occur in the matrix—at all scales. This is the essence of ecosystem management.

Thus, ecosystem management is not about bigness—it's about process; it's about maintaining a diverse, functioning ecosystem with an intact set of complex interactions. In developing this concept, David Mladenoff and John Pastor, two ecosystem ecologists who have made careers of studying process, take their experience in the northern hardwood and conifer forest and apply it to management. In expanding on Franklin's holistic perspective, they explore the potential for restoring important processes to the managed landscape.

Mladenoff and Pastor make the important point that traditional forestry is based on managing for an idealized state (stand structure, stocking, etc.). They argue that an ecosystem approach requires managing for process, and they underscore Franklin's point that this must embrace multiple scales. They introduce a new term to express a new way of thinking about management, *dynamic landscape heterogeneity*, which they believe captures the essential notions of process, scale, and pattern. Finally, to back up the theory, they recommend some silvicultural prescriptions that emulate the characteristic dynamics of the northern forest.

Mladenoff and Pastor benefit from the existence of a well-considered ecological land classification that reflects the natural community pattern. The dynamic history of the ecosystem is recorded in the landscape pattern and stand structure. Landscapes elsewhere in the United States are not so informative.

In the South, pine is king, and forest land traditionally has been classified in terms of its suitability for growing pine. There, before any progress can be made toward ecosystem management, a system of land classification that reflects ecological suitability must be developed. Steve Jones and Tom Lloyd discuss some of the obstacles to ecosystem management in the South and propose a new way of assessing ecological communities. Landscape ecosystem classification lays the groundwork for an ecological approach to forestry in that crucial region.

The South is not the only region where land-use history presents an obstacle to ecosystem management. In his chapter, George Parker describes the complex land-use patterns of the Central Hardwood region. There, the landscape is dominated by private ownerships without large tracts of contiguous forest land. Most stands originated after human disturbance and, hence, are young and relatively small, yielding a highly fragmented forest. Current land use precludes the restoration of presettlement, landscape-level disturbance regimes. Nevertheless, there are ways in which forest practices can be altered to maintain biodiversity. Parker recommends mixing longer rotations, uneven-aged management, corridors, and ownership consolidation into the traditional set of practices to ameliorate the ecosystem strains imposed by the current land-use pattern.

The situation is somewhat different in the inland West. There, forest contiguity does not limit ecosystem management. Instead, ecosystem management is limited by societal values—fear of large fires and demand for raw materials. The need to integrate societal values into forestry decision making is the foundation of Robert Pfister's discussion of ecosystem management. He recognizes that ecosystem management is not an ideal handed down from "on high." It must be chosen by society. The science of ecology can provide information to the decision-making process, but, ultimately, society must choose whether or not to manage sustainably.

Pfister adds to the discussion of Jones and Lloyd that ecosystem management requires knowledge not only of the current state of the forest but also of the potential for various future states. He argues, as do the others before him, that we must come to accept change and manage for it. Managing for desired future conditions requires the ability to predict the likely outcome of various current states. The recent maturation of simulation models and geographic information systems has dramatically improved our ability to assess the long-term outcome of management actions. Ecosystem management is poised to take advantage of this new technology.

Finally, John Gordon offers his "idiosyncratic" view of ecosystem management. As chairman of the Committee on Forestry Research of the National Research Council and a member of the Scientific Panel on Late Successional Forest Ecosystems, he brings both a national and a regional perspective to the discussion. Based on this experience, Gordon recommends that ecosystem management hold to five simple principles: (1) manage where you are, (2) manage with people in mind, (3) manage across boundaries, (4) manage based on mechanisms, and (5) manage without externalities. These economic principles provide the social context in which to begin to work with nature rather than battle against it.

The one constant through this entire section is the nonconstant: change. No matter what forest is being considered, change is inevitable. Ecosystem management, perhaps more than anything else, is about managing for change. Whether it is Franklin's New Forestry, Mladenoff and Pastor's dynamic landscape heterogeneity, Jones and Lloyd's landscape ecosystem classification, or Pfister's dynamic landscape ecosystem management, all of these concepts converge on a single point: managing ecosystems will require recognizing forests as dynamic units functioning at a multitude of spatial and temporal scales. Ecosystem management means restoring process to the forest; this must occur at both the stand and the landscape scale and must consider time scales both shorter and longer than rotation age.

4

The Fundamentals of Ecosystem Management with Applications in the Pacific Northwest

Jerry F. Franklin
College of Forest Resources, University of Washington

SUSTAINABILITY DEFINED

Sustainability is fundamentally a human construct. Hence, any discussion of resource sustainability is strongly conditioned by human values and objectives, whether these emphasize their short-term utility, their value in providing options for future generations, or, in the case of species, their inherent right to exist.

Sustainability, by my definition, refers to maintenance of the *potential* for our land and water ecosystems to produce the same quantity and quality of goods and services in perpetuity. Potential is emphasized because it makes implicit the option to return to alternative conditions rather than a permanent loss of some capability. This concept of sustainability considers a broad range of goods and services. For example, it includes retaining the ecosystem's capacity to provide functional services, such as the regulation of streamflow and the minimization of erosional losses of nutrients and soil. It means an ability to, either currently or at some future time, provide habitat for the full array of organisms historically found on the site and, of course, the continuing capacity to provide the same quantity and quality of products for human consumption.

The basis for sustainability, in my view, lies in maintaining the physical and biological elements of productivity. I propose that sustainability is based on two principles:

1. preventing the degradation of the productive capacity of our lands and waters—no net loss of productivity; and
2. preventing the loss of genetic diversity, including species—no loss of genetic potential.

I make three observations about these principles. First, each has both an ecological and an ethical basis (i.e., although they are human constructs, they can be objectively defined in ecological terms). Second, principle 2, no net loss of genetic potential, is probably the most fundamental, since we can sometimes restore productive capacity to degraded ecosystems but have only very limited capacity to restore lost genetic potential. Third, no principle (in my view) is absolute or inviolate. There will be times when rational, even ecologically sensitive, human beings will violate either principle. But when such violations occur they would be done with society's full knowledge of the act and its consequences, not as a result of ignorance and not in secrecy. Sustainability absolutely should not be viewed exclusively or primarily in terms of the short-term production of specific commodities, such as sawlogs or trophy ungulates, although such concerns are an appropriate component of a sustainable forestry concept.

Assuming the above, sustainable forestry should obviously place a very high priority on practices that meet the dual standards of (1) no degradation of productive capacity and (2) no loss of genetic diversity. Again, I urge a broad view of productivity—goods and services—and of the spatial and temporal scales to which it is to be applied, rather than a narrow construct. Many specific forest management practices can contribute to these dual objectives of maintaining productivity and genetic diversity. Books are full of them, and new approaches, often innovative mixtures of old practices and new ideas, emerge wherever work environments allow managers to be creative.

LIMITED VISIONS OF THE SUSTAINABILITY TASK

Unfortunately, many participants in the resource management debates have had limited visions of what needs to be done, whether they emphasize commodities or environmental values. The wood products forester has often seen productivity only in terms of soil productivity. Furthermore, interests have typically focused only on soil nutrients (or even just nitrogen). Physical soil properties (e.g., bulk density or aggregate structure) or, even more commonly, biotic properties (e.g., composition and structure of the communities of bacteria, soil animals, and fungi) have often been ignored. Furthermore, foresters have usually taken a short-term view of soil productivity rather than long-term views, such as the contribution of early-successional species with nitrogen-fixing symbionts or the importance of episodic depositions as a soil-forming process.

Environmentalists have, on the other hand, tended to have equally narrow emphases. In fact, almost everyone involved in the debates on the maintenance of biological diversity—environmentalist and utilitarian, manager and scientist—has focused on a very small set of genetic diversity—animals with backbones (or, even worse, only on the heroic megafauna) and vascular plants; the "verts" and the "vascs." Furthermore, participants in the debate have tended to focus on reserves as the primary approach to maintaining diversity. Typically, both sides want to isolate the biodiversity issue in some "preserves" or "set-asides," perhaps viewing that as a much simpler solution than trying to integrate their objectives and to collaborate in achieving them. Hence, arguments are generally on the required sizes or total area of reserves. Academic discussions of flows of organisms between reserves—connectivity—are similarly narrow, typically considering the pros and cons of corridors and largely ignoring the role of the landscape matrix.

Serious consideration of sustainability requires that we dramatically broaden our perspectives. All participants must reexamine old assumptions. There is a lot of history and personal bias present—among academic scientists, managers, corporate executives, agency heads, environmental activists, and so forth. We need to examine these assumptions in the light of evolving societal objectives for these forest lands and of our new knowledge about the functioning of ecosystems and landscapes.

THE ELEMENTS OF A SUSTAINABLE FORESTRY

What are some of the elements essential to my concept of sustainability? What are some of the perspectives that need to be added if we are to address it successfully? I propose five concepts for increased emphasis, recognizing that none is original with me: (1) thinking holistically—focus on ecosystems rather than species or products; (2) planning at large spatial scales—landscapes and regions; (3) recognizing the gradient of habitat maintenance—scales essential to species and processes range from square centimeters to thousands of hectares; (4) recognizing the multiple roles of the matrix in conserving diversity; and (5) recognizing that not all elements of diversity are equal and that more diversity is not necessarily better.

THINKING HOLISTICALLY

Thinking holistically about sustainability means approaching it on an ecosystem rather than a species or product basis. It means thinking about the full array of species, processes, and structures and about their interrelationships, which underlie the productive potential of a given site or ecosystem.

Ultimately, there really is no practical alternative to an ecosystem approach in conserving biological diversity, regardless of whether this perspective is a positive or negative contribution to our immediate personal and political agendas. Any other approach is doomed to strategic failure even if we can make it a successful tactic for a handful of species. Trying to conserve diversity on a species-by-species basis is going to exhaust our patience, pocketbooks, and the time and knowledge available.

An ecosystem approach is the only way that we are going to be able to conserve the overwhelming mass of existing diversity. The invertebrates, fungi, bacteria, and similar "lesser organisms" not only make up the vast majority of species but also carry out functions critical to the healthy functioning of our terrestrial ecosystems (i.e., they are essential, not decorative).

An ecosystem approach is also the only way that we are likely to conserve the organisms and processes found in poorly known or unknown habitats and subsystems. Ecological science in the last fifteen years has provided us with numerous examples of the richness of habitats previously unappreciated, such as the high canopies of the tropical and temperate rain forests (Irwin 1988; Schowalter 1989). Forest soils are now seen as subsystems that are intensely dynamic, diverse in species, and exhibiting spatially complex community patterns; they are also highly dependent upon copious, continuing energy supplies from vascular plants (Perry et al. 1989). The hyporheic zones associated with streams and rivers can be viewed as a newly discovered ecosystem, full of poorly known organisms yet functionally linked to the productivity of these streams and rivers (Franklin 1992). The ecosystems associated with hydrothermal vents in deep ocean areas (Myers and Anderson 1992) and with volcanoes, such as Mount St. Helens (John Baross, personal conversation, 1990) further instruct us that we have not yet even discovered the locale of all life and life processes, let alone catalogued it.

How, then, would we propose to deal with biological diversity using only, or even primarily, a species approach? We could not. Nor could we do without a species approach, given our objectives as human beings. Regardless of how much we can realistically do in habitat preservation, specific concern for the heroic megafauna will almost certainly be needed, but it should not obscure the larger task before us, if we truly intend to maintain a significant component of existing diversity, nor confuse us as to essential strategies.

PLANNING AT LARGER SPATIAL SCALES

In addition to thinking of ecosystems, sustainability requires that we think and plan at larger spatial scales—landscapes and regions. Again, in the long run it is the only way that we are going to be able to achieve our objectives. We need to have well-distributed reserves and plan for the flows of organisms between them. We have to think beyond the level of the individual landscape patch to its interactions with other landscape elements and the collective effects of our activities.

Planning at the landscape level is the only way that we are going to avoid undesirable, if not unacceptable, landscape dysfunction. Examples are conditions that we label as *cumulative effects* and *habitat fragmentation*. In the Pacific Northwest, large-scale clear-cutting affects streamflow, particularly peak flows associated with the regionally important rain-on-snow flood events (Harr 1986). As another example, the extent of erosional processes, such as landslides, is linked to the size of the road system, especially on steeper topography (Swanson and Dyrness 1975; Swanson et al. 1987). Dispersed patch clear-cutting, using patch sizes of 10 to 20 hectares, produces residual forest fragments that are subject throughout to the microclimatic influences of surrounding clear-cut areas (Chen 1991) and, in any case, have low viability (Franklin and Forman 1987).

Can anyone really believe that such landscape dysfunctions are nonexistent—the figments of environmentalists' imaginations? That alterations of a landscape cannot reach a sufficient scale to alter its ability to regulate aspects of the hydrological cycle or provide habitat for forest-dwelling organisms? I find it difficult to understand how some foresters can comprehend and accept the notion that one can alter a landscape to grow more wood, produce more water, or provide more edge for game species and, yet, cannot understand how such alterations might also produce large-scale (cumulative) negative effects on other resource values. Planning and acting with a landscape perspective is clearly one element in assuring sustainability.

Recognizing the Gradient of Habitat Maintenance

In conservation biology the focus of habitat maintenance has been on reserves—maintaining larger viable patches or areas of suitable habitat for species of interest. As mentioned earlier, this orientation to reserves has been equally true of the utilitarian and the environmentalist; differences are most commonly in how much to preserve, not the appropriateness of the strategy. Environmentalists have particularly emphasized the need for very large reserves to accommodate wide-ranging species and on wild and aesthetically pleasing landscapes; the "greater ecosystem" concepts that have been applied to the Yellowstone and North Cascade regions are reflections of these emphases. Academic conservation biologists have also been obsessed with reserves and, especially, the importance of reserve size and isolation.

Much of the focus on reserves is appropriate and understandable. Natural habitats, whole categories of ecosystems, are disappearing at a rapid rate as human populations and activities expand. Saving some pieces of habitat must have a high priority if we are to retain the species and processes dependent upon those habitats. Existing knowledge indicates a very high probability that there are such species and processes. For instance, there is only a limited amount of old-growth forest left in the Pacific Northwest, and, over at least the short term, it is our only source of habitat reserves. Hence, decisions about the amount and distribution of reserved late-successional forest habitat have high priority.

A comprehensive strategy for conservation of biological diversity and, hence, sustainability is almost certainly going to require areas— "reserves"—where human disruptions to natural processes, whether for extractive or other purposes, are minimized. The emerging evidence concerning forest reserves is that more and better-distributed reserves are going to be needed than we had suspected and that some of these are going to have to be quite large (Thomas et al. 1990).

However, reserves cannot be the only or even, perhaps, the primary strategy for maintaining biological diversity. There are many reasons, including the fact that we almost certainly could never create enough large, well-distributed preserves to do the job. Distribution is a particularly difficult issue, since the most productive parts of our temperate forest landscapes have been converted to human occupancy and use. Consider where our agricultural lands and tree farms are located, for example, in comparison with the locales for large preserves, such as Yellowstone and Alaska.

The issue of poor reserve distribution is an obvious one with the forest lands of the Pacific Northwest. Species diversity is highest on the most productive forest sites found at low elevations (Harris 1984). Such lands are overwhelmingly in private and state ownership. Hence, almost all existing and potential reserves of any size are on federal forest lands, which represent the less productive and less diverse habitats. This also makes a geographically well-distributed system of reserves unlikely.

At least as important to understanding the limitations of reserves is the recognition that the conservation of biological diversity requires habitat maintenance across a wide array of spatial scales. Habitat must be maintained on a continuum from the scale of meters to kilometers, from individual structures, such as snags or logs, to patches of old-growth forest and from special habitat conditions within commodity lands to preserves.

Many organisms will not be well served by a conservation strategy based on a few large reserves. In the Pacific Northwest, for example, large, widely spaced reserves may work for northern spotted owls and other organisms capable of dispersing large distances, but they are not likely to be adequate for tailed frogs, Olympic salamanders, or the numerous invertebrate species found in old-growth forests that are poor dispersers. Suitable habitat for such species needs to be maintained on a much finer spatial scale. And it is often possible to do this by maintaining structures, such as snags or downed logs, or protecting habitats, such as riparian zones and lake margins, within areas otherwise used predominantly for consumptive purposes, such as the production of wood fiber. In effect, reserves in the traditional sense are not needed to conserve many elements of diversity, but the maintenance of critical habitat is still essential. Unfortunately, many of our forestry and agricultural practices eliminate such habitats in pursuing efficient, short-term production of commodities.

It seems incumbent upon all interested parties, then, to recognize the critical role of the unreserved lands—the "seminatural matrix" (J. Brown, personal conversation, September 1990)—in maintaining biological diversity. Utilitarians must give up their notion that they can isolate their obligations with "set-asides." Conservationists and conservation biologists must back away from their nearly exclusive focus on reserves as the primary strategy for maintaining diversity. It is time to recognize the predominant role played by the landscape matrix.

RECOGNIZING THE MULTIPLE ROLES OF THE MATRIX

The landscape matrix, the dominant element within which a reserve system is embedded, plays at least three critical roles in conserving diversity: (1) providing habitat at smaller spatial scales, (2) increasing the effectiveness (buffering) of reserved areas, and (3) providing for connectivity. Role 1 of the matrix in providing habitat was discussed in the preceding paragraphs; within the temperate zone there are, in my opinion, more elements of diversity whose fate depends upon our treatment of the matrix than on reserves. Also, we may want to maintain well-distributed habitat for organisms, such as mycorrhizae-forming fungi or predatory invertebrates, within a commodity-oriented matrix for the direct benefits that they provide and not just to maintain biological diversity.

The importance of the matrix in influencing the effectiveness of reserved areas has been discussed by many authors, including Harris (1984). Reserves can achieve their objectives at smaller sizes when they are surrounded by patches that are structurally similar. Conversely, if the matrix that surrounds them is highly dissimilar, a larger reserved area will be necessary to provide the same conditions. As an example, Harris (1984) observed that a 25-hectare patch of old-growth forest surrounded by other late-successional forest would, if surrounded by clear-cuts, have to be 250 hectares to provide comparable conditions. Regardless of whether Harris was precisely correct in his selection of patch sizes, he was conceptually correct; it is very clear from studies of edge effects at old-growth and clear-cut boundaries that the effects of clear-cuts on environmental variables and biological processes extend deep (200 to 400 meters) into adjacent old-growth forest patches (Chen 1991; Chen, Franklin, and Spies 1992).

The role of the matrix in "connectivity"—particularly the flow of organisms and materials between reserves—is also in drastic need of attention. Again, the focus of the academic community has been on corridors; a large body of literature has accumulated debating the design and effectiveness of corridors. A general current consensus seems to be that corridors are important landscape elements in maintaining diversity and probably do facilitate the movement of some organisms, although this is not based primarily upon experimental evidence (Saunders and Hobbs 1991). Few biologists debate the importance of protecting riparian corridors. However, this consensus is based primarily on the need to protect values intrinsic to riverine and riparian habitats, including facilitating the movement of aquatic and riparian organisms. There is no consensus and little evidence that riparian corridors are effective at facilitating the movement of upland organisms, such as northern spotted owls.

Management of the matrix so as to facilitate connectivity between reserves and other isolated habitats is critical, then, once we recognize the ecological and practical limitations of corridors. In effect, we want to manage the matrix so as to make it less hostile to dispersing organisms or, in island biogeographic terms, to "make the sea between the habitat islands less deep." This could mean providing "haul-out areas" or stepping stones—small patches of suitable habitat within an otherwise unsuitable matrix. Maintaining within the matrix structural conditions that provide for protection from predators or hostile climatic conditions is another strategy to improve connectivity by matrix modification.

In the Pacific Northwest, the importance of matrix conditions to connectivity is increasingly recognized in developing strategies for conserving forest biodiversity. The development of the "50-11-40" recommendation as part of the Interagency Scientific Committee's plan for conserving the northern spotted owl is one example (Thomas et al. 1990); this rule proposes that the forest between habitat reserves be managed so as to maintain 50 percent of the area in trees 11 inches or larger in diameter and with a 40 percent crown cover. As is discussed below, in its recommendations for management of the forest matrix the Scientific Panel on Late-Successional Forest Ecosystems (1991) called for the maintenance of even more structural complexity.

The condition of the matrix is critical to overall connectivity. Management practices may produce very hostile conditions (i.e., deep seas full of sharks) or can be used to enhance dispersion (and in-place survival) of organisms. But the significance of the matrix must no longer be ignored because of an obsession with corridor-based strategies. Our investment in connectivity—dollars, land, timber, or whatever—must be designed to obtain maximum benefits; matrix modification and corridors need to be carefully compared along with other alternatives.

DIFFERENTIATING AMONG ELEMENTS OF BIOLOGICAL DIVERSITY

This fifth and last element in considering sustainable forestry seems obvious—that not all elements of diversity require equal attention and that "more diversity," as measured by simple indices, is not necessarily better. Although this concept seems obvious, many, including many forest managers, seem either not to understand this concept or to ignore its implications. For example, some elements of diversity that currently require greater attention in sustainable forestry are those associated with scarce or rapidly disappearing forest conditions. Old-growth and other late-successional forests fall into this category. In the Pacific Northwest there is no shortage of young Douglas fir forests and there is no threat to the survival of species that utilize clear-cut environments, from deer mice to fireweed. Populations of such organisms are healthy, and suitable habitat is in large supply. Species found in old-growth forests, on the other hand, require greater attention because suitable habitat has declined precipitously during the last century to the lowest levels that have probably been seen in several millennia.

The use of indices of biological diversity may be one factor that has interfered with broad recognition that different elements of diversity have different values or, at least, require differing amounts of attention. For example, indices to species diversity usually weigh species equally (i.e., they do not provide any weighting factors at all). More species means a higher index of biological diversity, regardless of whether the species is a dandelion or a condor. Hence, a logical inference is that more is better—and since clear-cuts often have higher diversity than do mature forests, biological diversity will be enhanced by more logging. A logical but not very useful conclusion!

Similar problems occur at the landscape level. A common index to landscape diversity will produce higher values for a landscape with many varied patch types than for a landscape with few patches. There are also important questions about the basis for differentiating patches into different conditions and the effectiveness of different patch sizes. Take, as an example, the traditional regulated forest landscape of the forester, such as sixty different age classes of young Douglas fir forest. Such a landscape might have high diversity based upon some mathematical index— probably much higher than a typical old-growth landscape in the Douglas fir region. But such a landscape has low diversity in terms of many organisms and ecological processes that sense the entire landscape as consisting of only two or three conditions. Again, the qualitative and subjective elements in making judgments about landscape diversity should be obvious, yet foresters persist in asserting that a landscape with sixty patches of loblolly pine one to sixty years old is a more diverse and, therefore, more desirable landscape than one with fewer, varied patches of natural forest.

To summarize, not all elements of diversity require equal or comparable attention at either the species or the landscape level, mathematical indices notwithstanding. More is not necessarily better!

EFFORTS TO APPLY THESE CONCEPTS IN THE PACIFIC NORTHWEST

These concepts are emerging as major elements in efforts to develop a sustainable forestry in the Pacific Northwest. Indeed, they represent products of, as well as contributions to, the progression of efforts to resolve forest resource conflicts in the region. Contributing activities include the development of forest plans, the development of New Forestry concepts, efforts to develop a consensus on aspects of forest resource management on nonfederal lands (the Sustainable Forestry Roundtable), the development of "A Conservation Strategy for the Northern Spotted Owl" (Thomas et al. 1990), and the report of the Scientific Panel on Late-Successional Forest Ecosystems (1991).

THE NATIONAL FOREST PLANNING PROCESS

The national forest planning process has contributed significantly to the inclusion of sustainability concepts in management of the national forests. The process was directed by the National Forest Management Act and required, for each forest, the development of plans that incorporated interdisciplinary approaches and public involvement. The act also required the Forest Service to maintain biological diversity on the national forests. Criticisms of the planning process have been numerous and often justified; it has been a major learning process and took far more time and money than anyone expected.

The development of interdisciplinary perspectives and attempts at a comprehensive view of a national forest's resources are major contributions from the planning process. For the first time, the Forest Service began to address all resources on the same systematic basis that had previously been accorded only timber management planning. Even though of uneven effectiveness, the development and use of interdisciplinary teams gave a regular voice to other resource values, such as soils, wildlife, and water. Factual information on all resources and the effects of various alternatives was presented more systematically, sometimes even when this information did not support the selected alternative; this process was speeded along as outside groups became more proficient at identifying errors of omission and commission in the draft plans. Even some of the failures of the planning process have been contributions, most notably the recognized need for spatial information in both planning and implementing activities.

The task of creating a sustainable multiresource management program for the national forests would be much more difficult without the development of the interdisciplinary work force, the experience of the planning effort, and the associated exponential expansion in knowledge and analytic capabilities.

New Forestry Concepts

Ecological research programs in the Pacific Northwest have made major contributions both in the identification of management problems associated with traditional forestry practices and in developing solutions for these problems. The term *New Forestry* has been used to identify the concept of using ecological principles to integrate better environmental and commodity values at the stand and landscape levels (Hopwood 1991; Franklin 1992; Hansen et al. 1991).

All of the elements of sustainable forestry mentioned earlier in this chapter are basic components of the New Forestry concept. At the stand level, New Forestry calls for the creation and maintenance of managed forest stands that have higher levels of compositional and (especially) structural diversity than have been characteristic of intensively managed forest stands. Such approaches obviously recognize the need to maintain habitat or niches at smaller spatial scales, modifying the landscape matrix so that it can better fulfill its multiple roles.

The importance of landscape-level phenomena and the need to plan activities over much larger spatial and longer temporal scales than has been traditional in forestry is emphasized in New Forestry. Developing a systematic approach to reserved areas so as to assure that their number, size, and geographic distribution are sufficient to achieve conservation objectives is one example. Planning appropriate protection for riparian zones associated with streams and rivers is another. The distribution and size of managed patches (e.g., harvest areas) is a particularly difficult issue (Franklin 1992). While an extensive and often noisy dialogue is occurring about the merits and demerits of New Forestry, extensive on-the-ground application of the concepts is already occurring on many forest owner-ships.

THE SUSTAINABLE FORESTRY ROUNDTABLE

The Sustainable Forestry Roundtable (SFR) was an attempt to develop a consensus among various forest interest groups regarding aspects of the management of state and private forest lands in western Washington state. The SFR followed the successful Timber-Fish-Wildlife agreement, which focused primarily upon the effects of timber harvest activities on stream quality and fisheries. Although ultimately the SFR agreement was not endorsed by all participants and, therefore, fell short of its goal, it did produce a consensus on several important issues.

First, all participants agreed that late-successional forest and riparian ecosystems and associated organisms were the elements of diversity of primary concern. From representatives of environmental organizations to large industrial landowners, all could agree that these were the sensi-tive and/or increasingly rare elements of diversity that needed to be targeted.

Second, the industrial forest landowners and state forest land managers agreed to dedicate 10 percent of their lands to the maintenance of riparian and late-successional forest ecosystems in return for a period of freedom from court challenges and more restrictive legislation. The owners also agreed to allow ecologists to participate in the selection of the dedicated areas and to leave some standing green trees, snags, and downed logs on harvested lands.

The consensus on priorities in conservation and the acceptance by industrial forest landowners of responsibilities in attaining environmental objectives are important steps in developing a regional strategy for sustainable forestry.

THE INTERAGENCY SCIENTIFIC COMMITTEE

The Interagency Scientific Committee to Address the Conservation of the Northern Spotted Owl (ISC) was created by four federal agencies to develop a scientifically credible conservation strategy for the northern spotted owl. It completed and published its report, "A Conservation Strategy for the Northern Spotted Owl," in 1990 (Thomas et al. 1990). This effort was unique in many aspects, including its exclusive use of a scientific team and criteria in addressing a major forest conservation issue. Two important areas where the report broke new ground were its proposals on the size and distribution of reserves and on modified treatment of the matrix.

The conservation strategy developed by the ISC calls for protection of large blocks of suitable habitat, which are called *habitat conservation areas* (HCAs), in place of an earlier proposal for a system of one- to three-pair spotted owl habitat areas (Thomas et al. 1990). The HCAs are delineated and mapped so as to include a minimum of twenty pairs of owls in each and with a maximum distance between HCAs of 12 miles; these dimensions are based upon scientific criteria. Over two million hectares of federal forest land are included in the HCAs, and the ISC recommends no harvest on HCAs for at least an interim period.

Modified management of the forest matrix between the HCAs is the second major innovative proposal of the ISC. The issue is facilitating successful dispersal of northern spotted owls between HCAs; mortality of dispersing birds is typically very high in cutover areas. Corridors between the forest reserves were considered and rejected in favor of a matrix management strategy. The ISC recommends that at least 50 percent of the forest land base outside HCAs be maintained in stands of timber with an average diameter at breast height of 11 inches or greater and at least 40 percent canopy closure; this recommendation is known as the *50-11-40 rule*.

The recommendations for the reserves and matrix by the ISC represent major intellectual and practical contributions to the problem of maintaining biological diversity in Pacific Northwest forests. The reader is again reminded, however, that the ISC was directed to address only the northern spotted owl; strategies for the protection of old-growth forests and other species associated with late-successional forest conditions were not considered.

THE SCIENTIFIC PANEL ON LATE-SUCCESSIONAL FOREST ECOSYSTEMS

In May 1991 several committees of the U.S. House of Representatives impaneled a small group of scientists to develop alternatives for addressing the full array of environmental issues associated with late-successional forests on federal timber lands affected by the northern spotted owl (Scientific Panel on Late-Successional Forest Ecosystems 1991). The charges to the panel included mapping and ecologically grading all of the late-successional and old-growth (LS/OG) forests, developing a range of alternatives for protection of LS/OG ecosystems and associated organisms—including northern spotted owls, marbled murrelets, and sensitive fish stocks—and conducting a biological risk analysis and an economic evaluation of each alternative.

The scientific panel completed its activity and submitted its final report in October 1991. Fourteen major alternatives involving various levels of LS/OG preservation and—for ten of the alternatives—three levels of matrix management were presented and analyzed. The analyses included biological risk ratings (the probability that specific organisms or ecosystems will survive the next century) and calculations of sustainable allowable timber cut and associated employment and income in the region.

The recommendations of the Scientific Panel on Late-Successional Forest Ecosystems (often referred to as the Gang of Four) were partially built upon but go beyond the ISC report in developing a regional strategy for conserving biological diversity. As in the case of the ISC report, the panel took a regional approach in developing alternative systems of LS/OG areas. Large blocks of thousands of hectares were identified rather than a fine-scale pattern of reserves involving tens or hundreds of hectares.

Developing an ecosystem approach rather than one based upon a single species represents a significant advance in addressing biodiversity on federal forest lands in the Pacific Northwest. In contrast to the ISC (which was directed to consider only the owl), the scientific panel addressed late-successional forest ecosystems and all of the organisms believed to be associated with them, including habitat for sensitive fish stocks. A comprehensive plan of this type allows much greater efficiencies than are possible by addressing species and issues on an individual basis.

The recommendations of the scientific panel also recognized the critical role of the forest matrix, the lands surrounding any system of LS/OG reserves. As with the ISC, altering management of the matrix was viewed as a strategy superior to delineation of management corridors in providing both for movement of organisms between reserves and for conservation of organisms and processes within the matrix. The 50-11-40 rule of the ISC was viewed as inadequate for the larger, multiple-species objectives of the scientific panel, however; for example, the ISC rule did not require the maintenance of large live trees, snags, and downed logs in the matrix. Hence, the scientific panel developed a minimal matrix management recommendation that prescribes both 50-11-40 and the retention of at least fifteen large live trees, five large snags, and five large downed logs on each hectare.

The development, presentation, and assessment (both ecological and economic) of multiple alternatives is an additional original contribution of the Scientific Panel on Late-Successional Forest Ecosystems (1991). Instead of providing a single solution, the panel provided thirty-four alternatives ranging from continued high levels of timber harvest (comparable to those during the 1980s) to the protection of essentially all remaining LS/OG forests. The ecological and economic costs and benefits of each alternative are presented, but no preference is indicated. Having presented the factual information, the panel (appropriately) left it to societal institutions (e.g., the U.S. Congress) to select among the alternatives.

CONCLUSION

Visions of sustainability in forest resources have been far too limited. The focus has typically been on products, whether of wood fiber or grizzly bears. Knowledge basic to understanding and managing for sustainability has been unavailable. The biased and dogmatic approaches of both resource managers and academic scientists have contributed to the lack of vision and the failure to develop essential information and innovative approaches.

All parties interested in sustainability need to leave their crumbling towers and take a fresh look at their objectives. Much broader views are required than either environmentalists or utilitarians have previously demonstrated. The linkages between sustained productivity and maintenance of biological diversity must be recognized. The critical role of the seminatural matrix in maintaining diversity must also be recognized, leaving behind the notion that biodiversity is primarily a set-aside or reserve issue.

Finally, we need to ensure that all of the critical information is available to all affected parties. The time of secretive or privileged approaches is past—and public oversight and overflight make such stealth technologically obsolete anyway. All interested parties must have access to the same information bases, most particularly including geographic information system data bases and models. All of society has a stake in sustainability, and the more information that is available, the more intelligent our decisions can be.

REFERENCES

Chen, J. 1991. Microclimatic pattern and basic biological responses at old-growth Douglas-fir forest edges. Ph.D. diss., University of Washington, Seattle.

Chen, J., J. F. Franklin, and T. A. Spies. 1992. Vegetation responses to edge environments in old-growth Douglas-fir forests. *Ecological Applications* 2:387-396

Franklin, J. F. 1992. Scientific basis for new perspectives in forests and streams. In *Watershed management: Balancing sustainability and environmental change*, ed. R. J. Naiman, 25-72. New York: Springer-Verlag.

Franklin, J. F., and R. T. T. Forman. 1987. Creating landscape patterns by forest cutting: Ecological consequences and principles. *Landscape Ecology* 1(1):5-18.

Hansen, A. J., T. A. Spies, F. J. Swanson, and J. L. Ohmann. 1991. Conserving biodiversity in managed forests. *BioScience* 41:382-392.

Harr, R. D. 1986. Effects of clearcutting on rain-on-snow runoff in western Oregon: A new look at old studies. *Water Resources Bulletin* 22:1095-1100.

Harris, L. D. 1984. *The fragmented forest*. Chicago: University of Chicago Press.

Hopwood, D. 1991. Principles and practices of new forestry: A guide for British Columbians. British Columbia Ministry of Forests land management rep. 71.

Irwin, T. L. 1988. The tropical forest canopy: The heart of biotic diversity. In *Biodiversity*, ed. E. O. Wilson, 123-129. Washington, DC: National Academy Press.

Myers, F. S., and A. Anderson. 1992. Microbes from 20,000 feet under the sea. *Science* 255:28-29.

Perry, D. A., M. P. Amaranthus, J. G. Borchers, S. L. Borchers, and R. E. Brainerd. 1989. Bootstrapping in ecosystems. *BioScience* 39:230-237.

Saunders, D. A., and R. J. Hobbs. 1991. The role of corridors in conservation: What do we know and where do we go? In *The role of corridors*, eds. D. A. Saunders and R. J. Hobbs, 421-427. Chipping Norton, Australia: Surrey Beatty & Sons Pty.

Schowalter, T. D. 1989. Canopy arthropod community structure and herbivory in old-growth and regenerating forests in western Oregon. *Canadian Journal of Forest Research* 19:318-322.

Scientific Panel on Late-Successional Forest Ecosystems. 1991. *Alternatives for management of late-successional forests of the Pacific Northwest: A report to the Agriculture Committee and the Merchant Marine and Fisheries Committee of the U.S. House of Representatives.* Corvallis, OR: Oregon State University College of Forestry.

Swanson, F. J., L. E. Benda, S. H. Duncan, G. E. Grant, W. F. Megahan, L. M. Reid, and R. R. Ziemer. 1987. Mass failures and other processes of sediment production in Pacific Northwest forest landscapes. In *Proceedings, streamside management: Forestry and fishery interactions symposium*, eds. E. O. Salo and T. W. Cundy, 9-38. Seattle: University of Washington Institute of Forest Resources.

Swanson, F. J., and C. T. Dyrness. 1975. Impact of clear-cutting and road construction on soil erosion by landslides in the western Cascade Range, Oregon. *Geology* 3:393-396.

Thomas, J. W., E. D. Forsman, J. B. Lint, E. C. Meslow, B. R. Noon, and J. Verner. 1990. A conservation strategy for the northern spotted owl. Report of the Interagency Scientific Committee to Address the Conservation of the Northern Spotted Owl. Portland, OR: USDA Forest Service Region 6.

5

Sustainable Forest Ecosystems in the Northern Hardwood and Conifer Forest Region: Concepts and Management

David J. Mladenoff and John Pastor
Natural Resources Research Institute, University of Minnesota

The northern hardwood and conifer region is transitional between the mixed hardwood forest to the south and the boreal forest to the north. This ecotonal region is a landscape of extreme contrasts in forest ecosystem properties. These ecosystem properties derive from four dominant influences: (1) the strong effects of different tree species on resources that control forest growth and productivity, particularly light and nutrients; (2) the varied scales and types of disturbances; (3) the strong control on species composition exerted by browsing mammals and insects; and (4) a diversity of reproductive strategies. These processes are linked in strong positive and negative feedbacks that cause the forested landscape, in a natural condition, to cycle at different temporal and spatial scales (Pastor and Mladenoff 1992).

Classic silvicultural management of these forests has focused on maintaining particular states without regard for the dynamic nature of these processes (Baskerville 1985, 1988). Accomplishing this required lowering of utilization standards as forests were highgraded. Ironically, the net result has been a simplification of the forest at different scales, altering the processes further and propelling the forests into states that rarely if ever existed naturally (Baskerville 1988). If we are to maintain the diversity and productive potential of these forests, silviculture must sustain the processes responsible for cyclic patterns of productivity and diversity at proper scales. It can no longer maintain static states at scales that are at variance with the natural scales of processes inherent in the landscape.

The purpose of this chapter is to review how the logging history of the northern hardwood and conifer forest has altered biodiversity at different scales, to review how processes derive from the interactions of organisms with the resources that sustain them, and to suggest silvicultural techniques that enhance these processes rather than work against them.

LOCATION AND DESCRIPTION

Geographically, the northern hardwood and conifer forest (NHCF) region extends from Minnesota and the western Great Lakes to New England and the Canadian maritime provinces; thus, it is also known as the Great Lakes–St. Lawrence forest (Nichols 1935; Braun 1950; Rowe 1972). This region is bounded on the north in Canada by the central boreal forest and on the south in the United States by the central hardwoods region (figure 5.1). This transitional zone is related to the interaction of Pacific, Continental, and Gulf maritime air masses and the position of the Arctic fronts (Arris and Eagleson 1989; Delcourt and Delcourt 1987; Kutzbach 1987; Bryson 1966). The NHCF zone ends where boreal forest meets the steppe in central North America, as is also the case on other Northern Hemisphere continents (Walter 1979; Eyre 1968). Within the North American NHCF region, the moderating influence of the Great Lakes provides microclimatic zones that allow the persistence of both southern extensions of the boreal forest and northern outliers of northern hardwoods (Pastor and Mladenoff 1992). Mountainous relief is important in the northeast, producing steeper environmental gradients and vegetational zonation (Oosting 1956; Bormann and Likens 1979a, 1979b), but even here the maritime climatic influence dominates (Raup 1941).

Taken as a whole, this transitional region has a great diversity of tree species, all of which may occur on modal sites. This is in contrast to Eurasia, which has only a fraction of the conifer species diversity of North America (Pastor and Mladenoff 1992). Although we tend to attribute the simple nature of European forests to their long-term, highly managed state, it may also be due to inherently simpler ecosystems. In particular, the North American transition zone is unique in having its own transitional species—eastern white pine (*Pinus strobus*), eastern hemlock (*Tsuga can-*

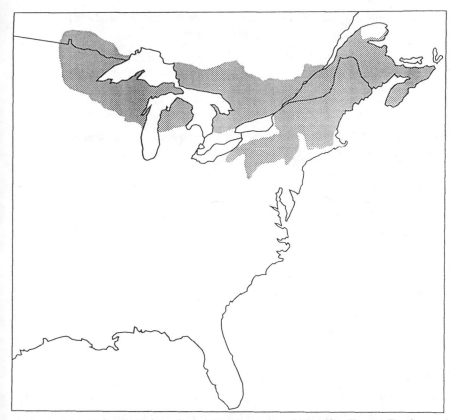

FIGURE 5.1. **Map of the Northern Hardwood and Conifer Forest Region.**
Modified from Bormann and Likens (1979) and Braun (1950).

adensis), and yellow birch (*Betula alleghaniensis*), which are characteristic of the northern hardwood zone itself (table 5.1) (Pastor and Mladenoff 1992; Little 1971).

In the largest part of this transitional region, particularly in successional forests, there is considerable mixing of characteristically boreal and northern hardwood species (Maycock and Curtis 1960; Bormann and Likens 1979a). The shared group of successional hardwood species between the spruce-fir and northern hardwood seres—aspen (*Populus tremuloides* and *P. grandidentata*), paper birch (*Betula papyrifera* and *B. populifolia*), and pin cherry (*Prunus pensylvanica*)—also adds considerably to the complexity of successional pathways and landscape heterogeneity within this region (Curtis 1959; Bormann and Likens 1979). Much of the region lacks

TABLE 5.1. **Major Tree Species of the Northern Hardwood and Conifer Region**

Species with Broader Southern Distributions, at Northern Range Limits		*Species with Broader Northern Distributions, at Southern Range Limits*	
Sugar maple	*Acer saccharum*	Balsam fir	*Abies balsamea*
Beech	*Fagus grandifolia*	White/red spruce	*Picea glauca/pubens*
Basswood	*Tilia americana*	Jack pine	*Pinus banksiana*

Characteristic Northern Hardwood and Conifer Region Species		*Common Early Successional Species*	
Yellow birch	*Betula alleghaniensis*	Aspen	*Populus tremuloides/ grandidentata*
Eastern hemlock	*Tsuga canadensis*	White birch	*Betula papyrifera/ populifolia*
White pine	*Pinus strobus*	Pin cherry	*Prunus pensylvanica*

the steep environmental gradients characteristic of more mountainous regions, but climatic variability is high (Hare and Thomas 1979; U.S. Department of Commerce 1968). This climatic variability is coupled with a large suite of tree species possessing varied reproductive and ecosystem properties. The result is a complex mosaic of stand types controlled by the more subtle interplay of species characteristics, site characteristics, and disturbances and producing considerable heterogeneity at the landscape and regional scales (figure 5.2) (Pastor and Mladenoff 1992).

It is particularly important to emphasize the transitional nature of this forest region and similar regions globally. The climatic and ecosystem variability associated with this region and its sensitivity to past and future climate change, along with the forest created by past use, will be major factors in the characteristics and long-term management of the regional landscape. The traditional concept of sustained-yield silviculture became widely applied during a particular period when forests in the NHCF region were largely in a regenerating phase. The assumption was that this state of constant productivity could be maintained and would meet all needs. Evolving knowledge of forest ecosystems and changing societal demands

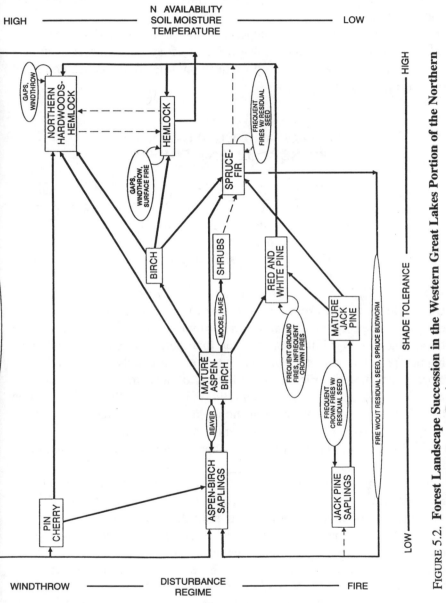

FIGURE 5.2. **Forest Landscape Succession in the Western Great Lakes Portion of the Northern Hardwood and Conifer Forest Region.**
Modified and expanded from Pastor and Mladenoff (1992).

suggest otherwise (Baskerville 1985, 1988). The silvicultural tools must be modified to manage ecosystem processes, and not states, at larger temporal and spatial scales on the forest landscape. It is within this context that we shall address further the conceptual changes, goals, and management strategies necessary to move toward forest ecosystems that are ecologically sustainable and that also provide economic commodities and meet other societal needs.

HISTORY AND CHANGE IN THE PRESETTLEMENT FOREST

The history of the NHCF has been one of highgrading and consequently progressive lowering of utilization standards; this sequence of events has placed the forests of the region in states for which there are no known natural analogues and whose processes are only vaguely understood (Baskerville 1985, 1988). Therefore, we shall begin by summarizing the forest history of the region.

After the Civil War, industrialization caused lumber production in the United States to soar from 8 billion board feet to 45 billion board feet at the turn of the century. The northern hardwoods and conifer region, from New England to Minnesota, was at that period the major source of lumber production in the United States and arguably experienced the most rapid and destructive exploitation of any forest region on the continent (Williams 1989; Flader 1983). In the Northeast, which was largely settled by Europeans one hundred to two hundred years earlier than were areas farther west, this catastrophic removal of the forest was primarily centered in interior and higher elevation areas (Bormann and Likens 1979a).

The changes resulting from this forest removal dramatically altered regional ecosystems at all scales. Forest species composition and stand structure, ecosystem processes, and wildlife habitat were all changed (Flader 1983). In particular, the combination of destructive logging practices and uncontrolled, repeated fires made recovery to any semblance of the presettlement state impossible (figure 5.3) (Bourdo 1983; Stearns 1990). The entire region was altered at both the forest stand (species composition and structure) and landscape scales (the relations of size and location of various forest ecosystems) (Mladenoff et al. 1993; Pastor and Broschart 1990).

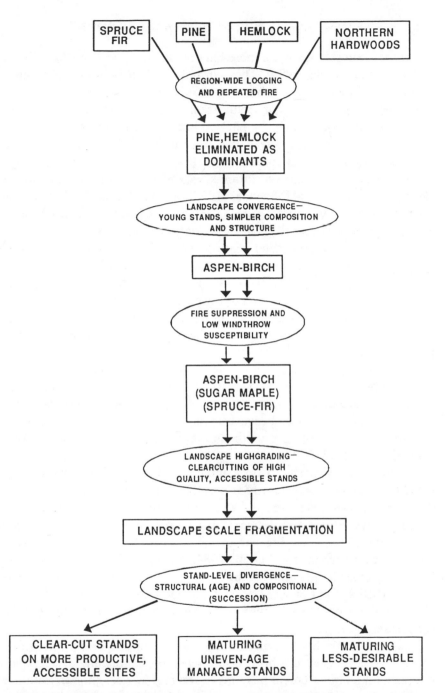

FIGURE 5.3. **Forest Landscape Conditions of the Twentieth Century in the NHCF Region during Which the Sustained Yield Paradigm of Forest Management Has Prevailed: Interaction of Landscape and Stand-level Processes and Effects.**

This new forest landscape differed in several key ways from the original forests (Stearns 1986, 1990). These differences not only dictated subsequent management and use, but also influenced the way in which forest states and processes were conceptualized. This in turn influenced the way forest management has been taught and implemented during this century.

The new successional forest that arose during this century has been dominated by deciduous species, with fewer species overall. Although conifers remain important on certain sites, especially northern and higher elevation areas of boreal influence, the regional landscape that had largely been dominated by conifers since glaciation has been radically altered both compositionally and structurally. Compositionally, this conifer dominance for nearly ten thousand years progressed from spruce to pine and hemlock, with spruce remaining in boreal areas (Davis 1981; Delcourt and Delcourt 1987; Webb 1974). Today the pine and hemlock regions are most altered; these species have been largely eliminated as regional dominants. The alteration of spruce-fir areas is largely successional, with greater prospects for regeneration of conifers where aspen and birch now predominate in those areas.

The highest elevation areas in the northeast are a separate case. Fir regeneration occurs sequentially (Sprugel 1976), but regeneration of high-elevation spruce may not occur because of global change (Solomon 1986; Pastor and Post 1988) or because spruce litter depresses soil nitrogen availability, thus limiting growth and reproduction (Pastor et al. 1987; Hamburg and Cogbill 1988).

The former northern hardwood- and hemlock-dominated areas are also those most changed structurally (figure 5.4, top). These forests had trees that were the most long-lived and were of the greatest stature. Disturbance regimes were also much longer during presettlement times (>1,000 years), far exceeding the lifespan of the dominants (250 to 500 years) in the Midwest (Canham and Loucks 1984; Frelich and Lorimer 1991) and becoming even longer moving to the northeast (Bormann and Likens 1979b). Frelich and Lorimer (1991) estimated that over 85 percent of the forests of the upper Great Lakes were in mature or old-growth stages at the time of settlement.

In contrast, the boreal forest, with more frequent disturbances and shorter lived species, has been less altered by harvesting. Disturbance intervals in the more boreal portions of the region were shorter, with fire intervals of one hundred years in northern Minnesota and Canada (Heinselman 1973) and frequent wave-regeneration of fir in the higher moun-

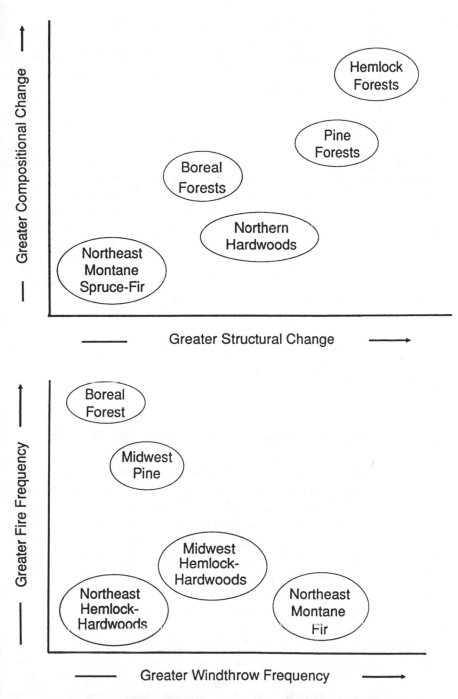

FIGURE 5.4. **Comparative Magnitude of Structural and Compositional Changes (*top*) and Relative Frequency of Stand-initiating Disturbance (*bottom*) among the Major Forest Types within the Northern Hardwood and Conifer Forest Region.**

tains of the Northeast (figure 5.4, bottom) (Sprugel 1976). Episodic infestations of spruce budworm (*Choristoneura fumiferana*) also cause extensive mortality in fir-spruce stands (Morris 1963). Therefore, the forests that are most altered compositionally and structurally by logging are the northern hardwood/conifer forests that had the largest, most long-lived tree species and that had the longest natural disturbance intervals. The net result of logging is a homogenization of stand ages and disturbance intervals across the boreal-northern hardwood/conifer region.

Also during this century, populations of forest animals that have key roles in the natural ecosystems have been severely reduced or eliminated, either through habitat change or directly through harvesting. Successional forests dramatically increased habitat for white-tailed deer (*Odocoileus virginianus*), while human hunting and trapping pressure eliminated or drastically reduced moose (*Alces alces*), other large ungulates, beaver (*Castor canadensis*), and carnivores, including the timber wolf (*Canis lupus*). Although it has long been known that browsing mammals such as deer can have a large direct effect on forests (Stearns 1949; Stoeckeler, Strothmann, and Krefting 1957), we are learning that large mammals induce other feedbacks within forest ecosystems and play a more complex role (Pastor and Mladenoff 1992).

CURRENT MANAGEMENT AND CONDITION

Logging of the original forests of the NHCF region has often been criticized in two major ways: (1) past practices caused uncontrolled, repeated severe fires and (2) stands were highgraded, removing only the largest, high-quality sawlog trees from each stand, leaving poorer trees behind. Today, the forest landscape is either recut or entering more mature stages after the resetting of succession by logging and fire on a regional scale 75 to 150 years ago (figure 5.3). There is, therefore, a growing situation on the landscape where the new forest—within a type relatively homogeneous in age, composition, and structure—is diverging into either very young, recently clear-cut stands or maturing, even-aged stands (e.g., as shown for aspen in Minnesota, figure 5.5) (Gephart et al. 1990). This has come about as the second growth forest reached harvestable age. Pulp has become the dominant timber market, and mills harvest by clear-cutting the highest quality, nearest stands first. Thus, the regional landscape is being high-

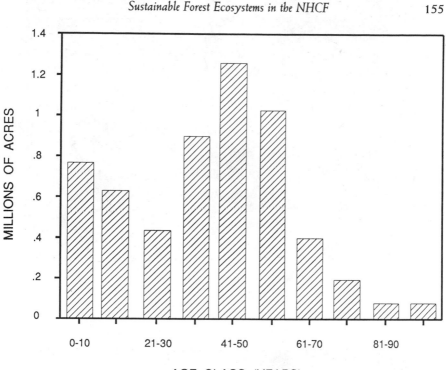

FIGURE 5.5. **Age Classes of Second-Growth Aspen Forest in Minnesota.**
From Gephart et al. (1990).

graded in much the same way that individual stands were highgraded for sawlogs in the past (Regier and Baskerville 1986). As in the past, this situation could be avoided by managing forest ecosystems in an integrated landscape context to the benefit of both ecological and economic sustainability. This will be addressed further in later sections.

Besides these effects on stand structure, the prevailing harvest pattern has changed landscape structure. The original forest landscape was a complex patch mosaic of differing forest types and ages. Past exploitation and recent cutting patterns have cumulatively simplified and fragmented the natural landscape by creating more, smaller, and simpler patches than existed originally (figure 5.6) (Mladenoff et al. 1993; Pastor and Broschart 1990). Today, the smaller forest patches contain stands

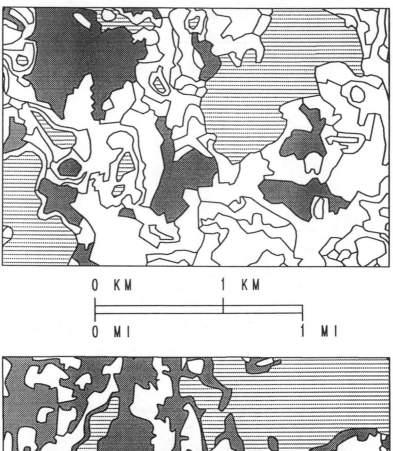

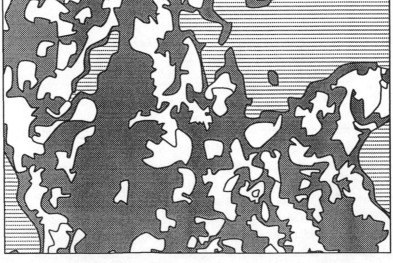

FIGURE 5.6. **Landscapes Illustrating Contrast in Forest Patch Size and Shape.** *The lower landscape is primary forest, largely old-growth hemlock-hardwood types, with dominant old-growth hemlock shaded; lakes are hatched. The upper landscape is second-growth managed forest (dominant northern hardwoods shaded) that originally had the same forest cover as the lower landscape. The two landscapes are located on the same landform and soils. The managed landscape is significantly simpler in patch size and shape.* Modified and redrawn from Mladenoff et al. (1993).

with fewer species maintained at truncated successional stages. Consequently, age class distribution and structural diversity are reduced (Frelich and Lorimer 1991).

DYNAMIC LANDSCAPE HETEROGENEITY—STATE VERSUS PROCESS MANAGEMENT

The application of traditional sustained-yield forest management during this century evolved when most eastern North American forests were in a similar, regenerating state. Commodity production and forest productivity and yield predictions were based on conditions of this short ecological period, characterized by simpler, homogeneous, largely even-aged forests with relatively constant rates of productivity over short periods early in succession, as in thirty- to eighty-year-old aspen stands (Pastor and Post 1986). The original widespread exploitation followed by repeated fire and then by equally unnatural fire suppression produced this great stand age and compositional uniformity (figure 5.3). This largely second growth landscape was also much less susceptible to windthrow, the other natural disturbance mode. One result was that the importance of catastrophic disturbance was underappreciated for some time into this century (Heinselman 1973; Loucks 1970; Canham and Loucks 1984). Similarly, the removal of native large mammals and the short period of observation led to an underappreciation of their role within the cycles of forest disturbance and regeneration (Naiman, Melillo, and Hobbie 1986; Naiman, Johnston, and Kelley 1988; Pastor et al. 1988).

The result was a paradigm based on an observation period that conveyed an impression of the forest as a constant state yielding relatively constant production. The prevalence of a similar, regenerating state in the forest landscape conveyed the impression that such a state could be maintained ("sustained") to provide a constant timber supply ("yield"). Classic silviculture developed in this simplified landscape with a paradigm that there is a maximum potential site productivity, which is fixed. The job of the forester is to assess potential productivity using devices such as site index and then manipulate the stand to bring actual productivity up to potential, which can theoretically be maintained. Forest models that have been developed to predict changes in timber supply are similarly based on this short, simple, and constant observational period (Belcher, Holdaway, and Brand 1982; Brand, Holdaway, and Shifley 1988). We are learning that this stage of the forest landscape is in fact unstable and an artifact of the conditions discussed above and the ecologically brief period of observation (figure 5.3) (Shugart, Crow, and Hett 1973; Stearns 1990, 1986).

The sustained-yield paradigm that dominated this century was based on several key assumptions: (1) potential forest productivity is constant, (2) utilization predictions can be based on managing the forest in a specific state to bring actual production up to potential, and (3) the observed state can be maintained indefinitely through management, without negative effects. With a full appreciation of the natural complexity of forested landscapes, these assumptions become untenable.

To develop sustainable forest ecosystem management, we must change from managing forest states to managing processes—from focusing on trees to focusing on ecosystems. As opposed to compartment and stand-level management, forest ecosystem management must be placed in a larger context we refer to as *dynamic landscape heterogeneity*. The intentional emphasis here is on both spatial and temporal change on the forest landscape due to varying processes and patterns. Processes appear linear and states appear constant only over a limited spatial and temporal field. We must emphasize the placing of forest landscape objectives in a larger context in both space and time. To do this, foresters must shift their emphasis from maintaining the forest in a given state to maintaining particular processes, even if the latter cause abrupt local changes in states.

This integration of management on the landscape scale also implies an integration of biodiversity maintenance objectives with harvesting. Reserves are a necessary part of such an integrated landscape, and it is likely that larger reserves are needed to maintain essential processes and provide both particular habitat as well as colonization nodes for ecosystem recovery. But biodiversity maintenance, as well as prevention of environmental degradation, must be seen as essential to overall sustainability, not merely as a set-aside issue (see chapter 4). Sustainability defined as maintaining long-term productive potential implies that all ecosystem components— the full array of biological diversity at species and genetic levels—must be maintained, both for their role in current productivity and for their potential role under changed environmental conditions or demands. This emphasis is the result of our increased understanding of the functioning of forest ecosystems at varied scales, as well as a beginning appreciation of potentially important species and processes that we do not yet understand in detail.

FOREST ECOSYSTEM PROCESSES AT VARIED SCALES

Recent important advances in forest ecosystem science have gone beyond mere description of conditions or states presumed to be static. The emphasis is now on greater understanding of linked processes, the positive and negative feedbacks between linked processes, and the relationship between processes at different temporal and spatial scales. Current understanding has replaced linear concepts of succession to stable climax stands with cyclic fluctuations of populations caused by linkages between processes (Wein and El-Bayoumi 1982; Pastor and Mladenoff 1992). Linkages between elements as they cycle between trees and soils impose nonlinear dynamics on forest growth and stand development (Flanagan and Van Cleve 1983; Pastor and Post 1986; Mladenoff 1987; Pastor and Mladenoff 1992). These nonlinear dynamics constrain the assumption of constant site productivity, implied by the use of devices such as site index, to short periods without major changes in species composition of the forest.

For example, the cycles of carbon and nitrogen are reciprocally linked because soil nitrogen availability controls net primary production, but the types of carbon compounds returned to the soil in litter in turn control nitrogen availability by determining the activity of soil microorganisms responsible for decomposition. This reciprocal linkage between these two element cycles is further linked to population dynamics and species composition of forests because different tree species respond differently to soil nitrogen availability and produce litter of different carbon chemistries. In particular, conifer production is saturated at relatively low levels of nitrogen availability (Mitchell and Chandler 1939; Weetman 1968; Stewart and Swan 1970; Van Cleve and Zasada 1976), and conifers produce litter that depresses soil nitrogen availability (Flanagan and Van Cleve 1983; Pastor et al. 1984; Mladenoff 1987). In contrast, hardwoods produce litter that is more easily decomposable and therefore enhances nitrogen availability (Melillo, Aber, and Muratore 1982; Pastor et al. 1984; Flanagan and Van Cleve 1983; Mladenoff 1987), but hardwoods require more nitrogen to sustain growth than do conifers (Mitchell and Chandler 1939).

Similarly, in northern hardwoods, nitrogen availability and especially nitrification vary at the single-tree scale, depending on the dominance of hemlock or deciduous species and the presence of gaps (Mladenoff 1987). Nitrification is low under an intact canopy where hemlock predominates, as would be expected from its low-quality litter (Pastor et al. 1984). With the creation of a small gap and with only a small proportion of sugar maple in the canopy (and litter), however, nitrification rises to levels equal to those on plots with only hardwoods (Mladenoff 1987). This situation may interact with light competition to create a mechanism whereby sugar maple is resisted under the intact hemlock canopy but its invasion is facilitated in gaps.

This reciprocal linkage between carbon and nitrogen cycles may in turn impose cyclic patterns of productivity in northern forests as hardwoods or conifers succeed one another (Wein and El-Bayoumi 1982; Pastor et al. 1987; Pastor and Mladenoff 1992). After aspen or birch initially invade a site, nitrogen availability and hence productivity are high because of the easily decomposable litter; succession to spruce or fir depresses nitrogen availability, in some cases leading to dieback, opening of the canopy, and reinvasion of aspen and birch (Pastor et al. 1987). Forest management will therefore alter potential site productivity to the extent that changes in species composition imposed by management also alter soil nutrient availability. The system is therefore not linear, as is implied by the assumption of constant site index, but highly nonlinear, a system in which foresters, tree species, and litter are interacting components.

Browsing mammals can influence the cycles of these elements and tree population dynamics by foraging selectively on hardwoods and avoiding most conifers (Pastor et al. 1988, 1993; Pastor and Naiman 1992). Although high population densities of browsers such as moose may depress productivity (McInnes et al. 1992), they may enhance the diversity of herbaceous species by opening the canopy. Therefore, populations of browsers may have different roles at different endpoints of their population cycles.

Insects also cause tree populations to fluctuate. For example, aspen and fir may alternate stand dominance as tent caterpillars (*Malacosoma disstria*) slow the growth of aspen and release understory fir (Duncan and Hodson 1958), but spruce budworm kills fir and allows aspen to reinvade (Morris 1963). The suppression of spruce budworm during the 1950s in Canada in the eastern portion of the NHCF region altered this stand dynamic to such an extent that the current maritime spruce-fir forest is in a unique state for which there is no historical precedent (Baskerville 1988).

By trying to manage for a particular state (spruce budworm–free monocultures of fir), foresters altered a cyclic process, which resulted not in the desired state but a state of reduced diversity and declining productivity about which we know little (Baskerville 1988).

We are learning that processes respond to many scales in these forests. Heterogeneity at within-stand scales of single treefall gaps, due to tree species composition and structure, influences patterns of nutrient cycling and species reproduction. The regeneration dynamics of plant species within a given stand seem to track this fine-grained, patchy resource heterogeneity. Some plant species, particularly those with bird-dispersed seeds and a buried seed bank strategy, accumulate higher densities of dormant seeds in small gaps, even before the gap is large enough for germination and establishment. But this may position these species to take advantage quickly of the higher resource availability in gaps as the opening enlarges and light, nitrogen, and moisture increase (Mladenoff 1987, 1990).

At the landscape scale in the former hemlock-hardwood-dominated areas, regional patch structure has been altered by the creation of a greater number of smaller and simpler forest patches of early successional forest types, in contrast to the old-growth hemlock-hardwood landscapes, which contained patch sizes an order of magnitude larger (Mladenoff et al. 1993; Pastor and Broschart 1990). Patches in the presettlement forest landscape were similarly much more complex in shape (figure 5.6). This was a landscape that came about through species responses to large, catastrophic windthrow disturbances in which various forest types exhibited characteristic adjacency relationships (e.g., old-growth upland hemlock contagious with lowland conifers) (Mladenoff et al. 1993; Pastor and Broschart 1990). This produced an interdigitated landscape structure conducive to the movement of animals and materials but with patch sizes that also still maintained substantial forest interior habitat. These relationships have been lost on the new managed forest landscapes (Mladenoff et al. 1993).

The current NHCF region landscape can be summarized in terms of the modification of states and processes at landscape- and stand-level scales brought on by historical changes (see figures 5.3 and 5.4). The regional landscape is fragmented but not primarily in the strict sense of formerly contiguous forest being reduced to permanently isolated, island fragments, as has occurred in more agricultural regions (Curtis 1956; Temple and

Cary 1988). Fragmentation in the largely forested NHCF region is characterized as a more qualitative fragmentation by simplification of landscape
patch structure and isolation of ecosystem-type patches. The balance of
regional species diversity is also reduced because of the loss of formerly
dominant mature forest types, habitats, and their associated species, when
compared with the presettlement condition. Reduction in natural disturbance regimes and modification of the role of large mammals have also
cumulatively affected landscape pattern and processes (Heinselman 1973;
Canham and Loucks 1984; Bormann and Likens 1979a; Pastor and
Mladenoff 1992).

At the stand level, changes are largely reductions in forest structure, tree
age classes, former species dominants, and associated habitats for plants
and animals (Stearns 1990; Frelich and Lorimer 1991; Mladenoff and
Stearns 1993). Many of these stand-level habitat modifications are closely
linked to changes in landscape-scale processes (e.g., Stearns 1990; Niemi
and Probst 1990).

The important point here is that cyclic patterns of succession and
productivity tied to the population dynamics of key tree and animal
species are a fundamental characteristic of northern ecosystems and have
implications at many scales. Rather than managing for a sustained level of
a particular target population, be it a timber species such as white pine or a
game animal such as moose, managers may consider sustaining the cyclic
nature of populations at the ecosystem level while maintaining a sustained
yield of a target population at the regional level. By thus modifying the
goals of management to the scales at which the processes work, we may
avoid some of the pitfalls of trying to apply the concepts of sustained yield
and multiple use at scales that are at variance with important ecosystem
processes.

DYNAMIC LANDSCAPE HETEROGENEITY AND THE
MANAGED FOREST

It is becoming increasingly recognized that managing sustainable forest
ecosystems requires viewing management objectives at larger spatial
scales (Mladenoff and Host 1993). The need to consider longer time scales
as well, to accommodate a broader range of objectives, has become
recognized more slowly. However, there are additional institutional levels

at which change must occur. Applying concepts of dynamic landscape heterogeneity to forest landscape ecosystems requires changes at the policy and planning levels, where entire national forests or state forest ownerships are placed in broader frameworks. Direction for establishing landscape ecosystem management must come from the administrative policy and planning levels. This direction is necessary to (1) institutionalize broader concepts of management objectives and (2) direct the application of new knowledge of forest ecosystems and landscapes and of techniques to the accomplishment of these broader objectives.

Similarly, a broader temporal and spatial context must be recognized in the application of management tools, primarily silviculture. Accomplishing broader forest management objectives will not be done by simply applying established silvicultural tools. These tools will require creative modification and application. For example, where even-aged management remains desirable, applying standard current clear-cut methods at a larger scale to regenerate larger forest patches on the landscape will solve one problem but create several others. The silvicultural toolbox must be enlarged, with current methods sometimes combined and applied over a longer time, to meet more complex objectives that promote ecosystem sustainability at both landscape and stand levels and over longer time frames. We are fortunate to have a diversity of forest tree species with varying ecosystem and life history characteristics in the NHCF region. This fact allows the development and application of more creative silvicultural techniques that provide for long-term commodity production and the maintenance of biodiversity and long-term productive potential, thus buffering against climatic changes. Silvicultural objectives must be expanded from simply regenerating a single species and thus (we hope) maintaining the forest in a given state to maintaining the processes that sustain biological diversity and productive potential even if this causes alternating states. Many proposed modifications in silvicultural treatments, both even-aged and uneven-aged, suggest longer rotation ages. Because of past exploitation and more recent management decisions, such changes will result in a modified and more diverse product mix deriving from the landscape. Some of the larger scale consequences of these management changes are addressed under "Landscape-Scale Integration" (below).

STAND-LEVEL MANAGEMENT UNDER EVEN-AGED SILVICULTURE

In much of the NHCF region, attention is being directed toward increasing uneven-aged management of many forest types and implementing longer rotations in even-aged systems. Such changes will increase the product mix diversity deriving from managed landscapes as well as increase regional ecosystem, habitat, and species diversity. In much of the region, however, even-aged management will probably continue, and methods should be applied that can accomplish this and, at the same time, better maintain biological diversity and long-term potential productivity. By far the dominant forest type in this management category is aspen, which now dominates the region, particularly in the Lake States. Modifying stand-level silviculture for this forest type and similarly managed intolerant species could contribute significantly to broader ecosystem management goals that encompass both habitat diversity and long-term productive potential.

In much of the NHCF region, even-aged aspen stands, either maturing or recently regenerated, differ most from the natural stands they have replaced by a lack of dominant conifers. Although aspen stands are often thought of as being single-species stands, natural stands are often of mixed composition, including other hardwoods and conifers, particularly in later stages (Graham, Harrison, and Westell 1963). Clear-cutting on short rotations (thirty to fifty years) has traditionally had the objective of eliminating these minor associates as well as regenerating another aspen stand through sprouts.

However, several objectives related to biodiversity retention and the buffering of long-term productive potential could be accomplished by modifying this strategy to use partial cutting with an extended rotation and two stand entries. Such a strategy involves encouraging and manipulating the natural aspen-to-conifer successional sequence, particularly with fir, and, after disturbance (i.e., logging), a return to aspen (Graham, Harrison, and Westell 1963; Perala 1977; Brinkman and Roe 1975). The silvicultural technique is a combination of group selection and seed tree silviculture, except that patches of both aspen and conifers are selected at different stand entries. When many of these mixed stands originated, the aspen was established first and/or outgrew the conifers, which may be pine, spruce, or fir. The conifers are usually patchily distributed under the aspen. At the first entry, the mature aspen is removed either entirely or by group selection, but the conifers are deliberately retained and re-leased. In a patchy stand, aspen will regenerate by sprouting in openings between conifers. In a stand with a dense and uniform conifer under-story, group selection or low-density retention of conifers can be used to allow room for some aspen regeneration, maintaining the structural species mix and retaining the aspen clones within the stand through the period of conifer dominance. This combined group selection of aspen and retention of conifers adds structural and habitat diversity usually lacking in young stands.

At the end of the rotation, approximately 85 percent of the stand is cut. The majority of the conifers are removed, along with a second cutting of aspen. At this final entry, trees are retained in the stand for several purposes that meet both silvicultural and diversity goals. Conifers are retained as seed trees to regenerate a mixed stand, and the harvested aspen will again resprout a fast-growing aspen component.

An important modification of this technique is that additional trees beyond those needed as seed trees are retained in the partial cutting in patches and as single trees to make up the approximately 15 percent retained. The amount retained will vary with the conifer species and site and stand characteristics. These green leave trees and retained snags provide the "biological legacy" of the stand structure (Franklin et al. 1989). The microhabitat structure provided by this technique is essential for retaining species diversity (plants, vertebrates, invertebrates, and soil organisms) and providing for rapid recolonization of the stand (Franklin et al. 1989).

Such stands have direct silvicultural values as well. Maintaining such mixed stands and alternating aspen and conifer dominance results in a reduction of large-scale infestations of pests such as spruce budworm (Morris 1963) and tent caterpillar (Duncan and Hodson 1958) and of pathogens, which can be significant problems in intensively managed pure stands of both aspen and conifers in the NHCF region (Graham, Harrison, and Westell 1963; Baskerville 1988). At the same time, the alteration between aspen and conifer dominance mimics the limit cycle caused by these insects (Ahlgren and Ahlgren 1984) and exploits the opposite effects of aspen and conifers on the nitrogen cycle (Pastor, Gardner, and Post 1987). Also, because the site is never completely clear-cut, problems of soil erosion and loss of nutrients, soil organic matter, and important soil organisms are minimized, even on difficult sites.

Modifications of this strategy could be applied to pure aspen stands and to aspen with a minor component of other hardwoods, spruce, or pine, with the important emphasis on modifying the cutting regime to provide an admixture of longer lived species through subsequent rotations. At each aspen harvest, this process can be repeated, adding additional age classes of the secondary species. These can be retained through more than one aspen rotation, contributing both habitat and structure, such as snags and large downed wood, typically lacking in aspen stands. A proportion of these trees will also eventually provide large sawlogs.

Greater habitat and structural diversity could also be enhanced through modifying and extending the rotation of traditional shelterwood cutting typically applied to midtolerant and tolerant conifers and hardwoods, such as white pine, spruce, northern hardwoods, and to some extent red pine (Fisher and Terry 1920; Frothingham 1914; Tubbs and Metzger 1969; Ohmann et al. 1978). In traditional shelterwood cutting, the stand is typically opened with an initial cutting of 25 to 75 percent of basal area to provide light levels adequate for germination and establishment. The exact amount removed depends on the shade tolerance of the desired species and stand and site conditions. After regeneration is established, the overstory is removed in subsequent entries to release the established saplings. If the period between successive removals is less than 20 percent of the total rotation age, even-aged stands will develop (Smith 1962).

It would be possible to modify such procedures to obtain uneven-aged stands by retaining the overstory for longer rotations. The eventual subsequent cuttings could also retain green leave trees, as described for the mixed species partial cutting (above). This would accomplish several things. The frequency of stand entry would be reduced, lessening effects on the site. In terms of commodity production, larger, higher value logs would result, and the product diversity mix harvested from the landscape would increase. The large trees produced through the longer rotation would also maintain some old-growth structural and habitat values, such as canopy layering, nest trees, snag production, and coarse woody debris on the forest floor. This strategy would be particularly effective in the management of NHCF pines, which have been largely eliminated in older and larger classes (figures 5.3 and 5.4). Pines provide high-value, large sawtimber and important wildlife values because of their large stature, both in pine-dominated stands and as scattered individuals in stands of other types. They provide supercanopy nest trees for large raptors and large, persistent snags and downed logs.

A promising variation of shelterwood silviculture for mixed conifer-hardwood stands similar to the NHCF region has been developed in northeastern China. The forests of this region bear a striking resemblance to those of northeastern North America, having many congeneric species (Burger and Zhao 1988). Until very recently, the standard silvicultural practice has been to replace mixed species stands with monocultures of various species of larch (*Larix* spp). However, this has caused the decline of the more economically valuable species, particularly Korean pine (*Pinus koraensis*), which bears a striking ecological resemblance to eastern white pine (Burger and Zhao 1988).

Current management of these forests is a shelterwood-like technique in which 30 to 60 percent of the canopy is removed at twenty-five- to thirty-year intervals. Diameter criteria for removal are species specific and based on the culmination of mean annual increment, but the cuts are distributed across species to maintain the same ratios of basal area as in old-growth forests (J. Pastor, personal observation). The term for the new type of shelterwood cutting is *ecosystem regeneration*, as opposed to classic *single-species regeneration* (Zhao Shidong, personal communication). Given the ecological and taxonomic similarities between the regions, it is possible that such a technique could be applied here with some additional modifications for maintaining more large structural elements and older age classes. These could include significantly longer periods between stand entries and the retention of more older trees.

Modifying traditional even-aged forestry as described in the examples above would accomplish several important objectives. (1) The conifer component of the current, extensive aspen stands would be increased; (2) a greater number of older tree size classes would be created; (3) the amount of large living and dead woody structure and microhabitats would increase; and (4) substantial elements of the original regional forest, including large pines, would be restored in many stands. The precise details of a silvicultural treatment in a given situation will always depend on stand and site conditions and specific objectives. The preceding discussion is an initial suggestion of how current even-aged silvicultural practices may be modified to meet a broader set of objectives. We believe that many more such applications are possible.

STAND-LEVEL MANAGEMENT UNDER UNEVEN-AGED SILVICULTURE

In many ways, uneven-aged management, or various methods of selection cutting (retaining forest cover during harvest), requires the least modification to meet broader ecosystem objectives in the NHCF region. Although thought of as the logical low-impact alternative to clear-cutting, selection cutting as traditionally used is not a panacea. Selection cutting, applied uniformly to the landscape, can eventually reduce overall regional diversity, particularly in areas where former natural disturbance regimes were more frequent and intensive than the conditions produced by the implementation of selection cutting. Selection cutting can also severely fragment large areas of intact forest canopy, rendering the stand unsuitable for many forest interior species. Also, selection cutting can be inappropriate under certain site conditions, such as unstable soils and steep topography, because of the need for extensive roads and frequent harvest entries. These factors, combined with high precipitation, make selection cutting as problematic as clear-cutting in many parts of the Pacific Northwest (Franklin 1990); landscapes with similar limitations also exist within the NHCF region.

However, much of the NHCF region is ideal for modified selection systems. Topography is generally very moderate within most of the region. Also, even under old-growth conditions, the trees in the NHCF region do not approach the height or mass of those in the Pacific Northwest. Frequent selection cutting, therefore, is not as destructive to the remaining trees. Much of the region was originally characterized by long disturbance return intervals relative to tree age, so the NHCF region was dominated by late-successional forests. These factors all suggest that selective harvesting techniques, if modified in their application, can successfully be used in ways that enhance forest biodiversity and maintain long-term productive potential in this region. Selection systems are particularly appropriate to large portions of the NHCF region dominated or potentially dominated by northern hardwoods-hemlock and, in the northeast, by red spruce-northern hardwoods.

Selection cutting involves the removal of single or small groups of trees throughout a stand, potentially simulating natural gap dynamics in uneven-aged forests (Runkle 1982; Mladenoff 1987, 1990). Historically, selection cutting has been applied by removing half or more of the stand basal area, resulting in severe opening of the canopy and destruction of the original forest structure (Lorimer 1989). Cutting was also not distributed in a balanced way across all size classes, and most large trees were typically removed (Tubbs 1977). Selection was often poorly applied in a way that resulted in severe highgrading, removing all good growing stock as well as all mature trees (Tubbs 1977).

Applying selection systems in stands where broader ecosystem management objectives are to be met requires substantial modification of these original techniques, and some of these modified techniques have been developed (Arbogast 1957; Crow et al. 1981). Stands managed in such modified ways have the potential to maintain substantial aspects of old-growth and natural forest stand structure and function while providing long-term productivity and profitable commodity production (Pastor and Mladenoff 1993). The major difference is that, depending on objectives, stand basal area and coarse woody debris on the forest floor will be reduced to varying degrees from those of comparable natural stands.

More recent research and applications have shown that reducing cutting levels and maintaining a residual stand diameter class distribution that resembles the "reverse-J" or negative exponential of many natural stands can also maximize long-term forest growth and production (Lorimer 1989; Marquis 1978; Crow et al. 1981). Cutting levels are determined by the difference in current size class levels and the residual levels to be maintained.

Several additional modifications still must be made to this much improved system to meet noncommodity objectives. Traditionally, marking stands for harvest included discrimination against minor species and "undesirable" trees, usually defective or high-risk individuals. The new system also usually requires frequent stand entries (at intervals of five to ten years) (Lorimer 1989), but minor species would be deliberately retained, as would potential future snags, which would have been removed in the past. Less-frequent entry would allow for some mortality and maintenance of snags and downed logs and would be compatible with retaining more and older individuals than would be done with purely commodity objectives. Also, inasmuch as single-tree selection will gradually increase only sugar maple, larger multitree openings should be created. Along with cutting that discriminates against sugar maple, cutting during good seed years of the less-tolerant yellow birch, spruce, and other species along with larger gaps will favor regeneration of the minor species. Because of problems with windthrow and unbalanced structures, several decades would be required to bring maturing, mixed, even-aged stands under this type of management, but uneven-aged management of all-aged stands can begin immediately (Lorimer 1989; Erdmann 1986). With site- and stand-specific modifications, this system is potentially applicable over the largest part of the region on sites of average or better productivity. Many young, mixed, aspen-birch-northern hardwood stands that are currently clear-cut and managed for pulpwood or chip production could be converted to the all-aged management described here with benefits to both biodiversity and long-term ecosystem productivity.

LANDSCAPE-SCALE INTEGRATION

Classic silviculture dominates current forest management and has had great influence on the dynamics of forest stands. Additionally, classic silviculture directly affects forests at the landscape scale. In standard forestry, the forest is typically divided into management compartments that are subdivided into stands, consisting of tens of acres, to which particular silvicultural techniques are applied. Compartment and stand boundaries are generally considered to be fixed and independent, and each unit is assigned one appropriate silvicultural technique for one species, usually applied without great consideration to the larger context. Applying *dynamic landscape heterogeneity* in management requires that over longer time frames these boundaries should be considered fluid and that management techniques should be aggregated over larger areas to maintain landscape-scale patterns and processes. In the natural landscape, overlapping fires and other disturbances cause boundaries to be fluid (Heinselman 1973), but this dynamic process has been halted by compartment-based management.

General principles of forest landscape management to sustain biodiversity have been well elucidated (e.g., the need to incorporate corridors and buffers, to aggregate cuts over time to minimize induced edge and fragmentation, and to limit the amount and frequency of cutting adjacent to sensitive habitats and ecosystems such as old growth) (Franklin and Forman 1987; Noss 1987; Noss and Harris 1986; Harris 1984). We have also tried to show that sustainable ecosystem management requires that we maintain biological diversity both to sustain the health and productive potential of forest systems over long time frames and to buffer against long-term environmental change. Consideration of such concepts at larger spatial and temporal scales requires management integration at those scales (Mladenoff and Host 1993).

In the NHCF region, biodiversity can be conserved by maintaining natural landscapes with larger forest patches of more complex shapes than cutover forest landscapes (Mladenoff et al. 1993). Additionally, characteristic patterns of contagion (adjacency of different ecosystem types) occur in the natural landscape. These landscape matrix configurations maximize both forest interior and landscape interspersion at the same time, meeting specific wildlife habitat needs and facilitating movement (figure 5.6) (Mladenoff et al. 1993). Silviculturally, such patterns may be important in facilitating adequate seed dispersal and regeneration. These patterns can be restored to the managed landscape, maintaining a large degree of natural ecosystem functioning within the new managed matrix and still provide commodity production.

Constructing larger cutting units within even-aged forests can be accommodated but not by simply creating larger clear-cuts, which generate other problems. Implementing larger cuts on the landscape using partial cutting and the other modified, even-aged techniques described above can accomplish the broader objectives of sustainable ecosystem management. Cuts can be increased significantly in size without detrimental effect if made linear, to limit the greatest dimensional width, and if broken within the cutting unit with leave trees and buffer corridors. Modifying traditional short-rotation, aspen clear-cutting as we have proposed maintains greater structural elements and, where appropriate, allows an alternation and mixture of aspen with conifers that moves through the larger landscape over time. Integrating this type of management with longer rotation harvesting of longer lived and more tolerant conifers and northern hardwoods will provide a component of the larger landscape that cycles more slowly, both spatially and temporally, and provides critical ecosystem components of habitat and structure at stand and landscape levels. These diverse components of large landscapes, cycling at varying rates but undergoing change at larger time scales, constitute the application of dynamic landscape heterogeneity. In such landscapes, the processes of sustainable forest ecosystems and the patterns and uses they produce can be maintained.

ECONOMIC CONSIDERATIONS

Significantly modifying the management of the larger commodity forest landscape as we have described will change the forest product mix provided by the landscape. In general, larger tree products, such as sawlogs and veneer, will gradually increase. The species mix will also change, with a more diverse mixture of species, including conifers and hardwoods not now present in large quantities or sizes. Species managed on even-aged systems, such as aspen and conifers for pulp, will be reduced in acreage but will be maintained in a larger range of age/size classes.

Because highgrading and economic optimization have led to progressively smaller and younger forests, utilization standards have been reduced to a historical low (Regier and Baskerville 1986; Baskerville 1988). Species and tree sizes formerly considered useless are now being harvested at increasing rates. In the early part of this century, large old-growth hemlock were stripped of bark for tanning, and the logs were left to rot because they were considered inferior to pine (Corrigan 1976). Today such resources would be invaluable. Aspen, once considered little more than a weed (Graham, Harrison, and Westell 1963), became valued first for pulp and is now being utilized in ever younger age classes for chips used in composite wood products. These utilization levels, with whole stand removal at ever shorter rotations, may not be sustainable. Indeed, new mills using such technology are designed for short life and may contribute to rapid resource depletion down to unsustainable levels. Such utilization also requires an increasing input of technology, with greater energy inputs and pollution outputs, to maintain increasingly short-term profitability.

It may make more sense for both a sustainable forest and a long-term, sustainable forest economy to manage the accumulation of added value wood products in the forest rather than through expensive technology. Such changes in product mix as we have described will provide a more sustainable resource base of greater value, particularly to smaller sawlog and veneer mill owners whose mills are locally owned and operate on simpler technology, lower energy inputs, and lower pollution outputs. The most obvious change in the product mix of such forest management is the reduction in pulp acreage. However, Forest Service inventory results show that total wood growth in the United States still exceeds harvest by more than 30 percent (Bowyer 1992). In addition, increasing paper recycling to 50 percent—only 20 percent more than the current level—would reduce the need for virgin wood pulp by 25 percent (Bowyer 1992). These figures suggest that there is more than enough slack in the current supply system to implement a variety of changes over time to increase the product mix of managed forests as well as the age and size of many trees harvested.

CONCLUSION

Sustainable ecosystem management should be seen not merely as protecting additional old-growth reserves and providing landscape corridors, but also as a necessary change in managing the larger commodity forest matrix of the landscape. The concept of dynamic landscape heterogeneity emphasizes the need for long-term objectives of maintaining dynamic processes that change over larger spatial and temporal scales and not creating unstable, local forest states. The maintenance of species diversity, functional diversity, and processes through harvesting cycles is needed not only for wildlife, but also to sustain the long-term productive potential of forests and to provide a buffer against future uncertainty, such as climatic change. Although additional reserve areas are required to meet these needs and connecting corridors are a logical addition to the landscape, sustainable forest ecosystems require all of the management changes we have described to maintain biological diversity and forest productivity.

Significant informational needs remain in several areas. More research is required to understand further the links among landscape patterns, their temporal changes through time, and the processes that both sustain and require them. Long-term research that tests the success of alternative silvicultural techniques in creating and maintaining natural structure and processes at stand and landscape scales is also required. Additional new management techniques beyond those suggested here are also needed and should be a product of the research that continues to identify key forest processes and their links with structural elements and landscape pattern.

ACKNOWLEDGMENTS

We appreciate review comments from D. Perala, and we are grateful to P. Barnidge for preparing the figures. Various portions of the work reported here were supported by grants from the National Science Foundation (BSR-8906843), the Nature Conservancy of Wisconsin, and the USDA Forest Service, North Central Forest Experiment Station, Landscape Ecology Program. This is Contribution Number 97 of the Center for Water and the Environment.

REFERENCES

Ahlgren, C. E., and I. Ahlgren. 1984. *Lob trees in the wilderness.* Minneapolis: University of Minnesota Press.

Arbogast, C., Jr. 1957. Marking guides for northern hardwoods under the selection system. Research Paper LS-56. Detroit, USDA Forest Service, Lake States Forest Experiment Station.

Arris, L. L., and P. S. Eagleson. 1989. Evidence of a physiological basis for the boreal-deciduous forest ecotone in North America. *Vegetatio* 82:55-58.

Baskerville, G. 1985. Adaptive management wood availability and habitat availability. *Forestry Chronicle* 61:171-175.

Baskerville, G. L. 1988. Redevelopment of a degrading forest system. *Ambio* 17:314-322.

Belcher, D. W., M. R. Holdaway, and G. J. Brand. 1982. A description of STEMS, the stand and tree evaluation and modeling system. Gen. Tech. Rep. NC-79. St. Paul: USDA Forest Service, North Central Forest Experiment Station.

Bormann, F. H., and G. E. Likens. 1979a. *Pattern and process in a forested ecosystem.* New York: Springer-Verlag.

————. 1979b. Catastrophic disturbance and the steady state in northern hardwood forests. *American Scientist* 67:660-669.

Bourdo, E. A., Jr. 1983. The forest the settlers saw. In *The Great Lakes forest: An environmental and social history*, ed. S. L. Flader, 3-16. Minneapolis: University of Minnesota Press.

Bowyer, J. L. 1992. Responsible environmentalism: The ethical features of forest harvest on a global scale. *Forest Perspectives* 1:12-14.

Brand, G. J., M. R. Holdaway, and S. R. Shifley. 1988. A description of the TWIGS and STEMS individual-tree-based growth simulation models and their applications. In *Forest growth modelling and prediction*, eds. A. R. Ek, S. R. Shifley, and T. E. Burke, 2:950-960. Gen. Tech. Rep. NC-120. St. Paul: USDA Forest Service, North Central Forest Experiment Station.

Braun, E. L. 1950. *Deciduous forests of eastern North America.* Philadelphia: Blakiston Co.

Brinkman, K. A., and E. I. Roe. 1975. Quaking aspen: Silvics and management in the Lake States. USDA Agric. Handbk. 486. Washington, DC: USDA.

Bryson, R. A. 1966. Air masses, streamlines, and the boreal forest. *Geographical Bulletin* 8:228-269.

Burger, D., and S. Zhao. 1988. An introductory comparison of forest ecological conditions in northeast China and Ontario, Canada. *Forestry Chronicle* 64:105-115.

Canham, C. D., and O. L. Loucks. 1984. Catastrophic windthrow in the presettlement forests of Wisconsin. *Ecology* 65:803-809.

Corrigan, G. A. 1976. *Calked boots and cant hooks.* Park Falls, WI: MacGregor Litho.

Crow, T. R., R. D. Jacobs, R. R. Oberg, and C. H. Tubbs. 1981. Stocking and structure for maximum growth in sugar maple selection stands. Research Paper NC-199. St. Paul: USDA Forest Service, North Central Forest Experiment Station.

Curtis, J. T. 1956. The modification of mid-latitude grasslands and forests by man. In *Man's role in changing the face of the earth,* ed. W. L. Thomas, Jr., 721-736. Chicago: University of Chicago Press.

————. 1959. *The vegetation of Wisconsin.* Madison: University of Wisconsin Press.

Davis, M. B. 1981. Quaternary history and the stability of forest communities. In *Forest succession: Concepts and application,* eds. D. C. West, H. H. Shugart, and D. B. Botkin, 132-153. New York: Springer-Verlag.

Delcourt, P. A., and H. A. Delcourt. 1987. *Long-term forest dynamics of the temperate zone.* Ecological Studies 63. New York: Springer-Verlag.

Duncan, D. P., and A. C. Hodson. 1958. Influence of the forest tent caterpillar upon the aspen forests of Minnesota. *Forest Science* 4:71-93.

Erdmann, G. G. 1986. Developing quality in second-growth stands. In *Proceedings, Conference on the Northern Hardwood Resource: Management and Potential,* compilers, G. D. Mroz and D. D. Reed, 206-222. Houghton: Michigan Technological University, School of Forestry and Wood Products.

Eyre, S. R. 1968. *Vegetation and soils: A world picture.* 2d ed. London: Edward Arnold.

Fisher, R. T., and E. I. Terry. 1920. The management of second-growth white pine in central New England. *Journal of Forestry* 18:358-366.

Flader, S. L., ed. 1983. *The Great Lakes forest: An environmental and social history.* Minneapolis: University of Minnesota Press.

Flanagan, P. W., and K. Van Cleve. 1983. Nutrient cycling in relation to decomposition and organic matter quality in taiga ecosystems. *Canadian Journal of Forest Research* 13:795-817.

Franklin, J. F. 1990. Thoughts on applications of silvicultural systems under New Forestry. *Forest Watch* Jan/Feb:8-11.

Franklin, J. F., and R. T. T. Forman. 1987. Creating landscape patterns by forest cutting: Ecological consequences and principles. *Landscape Ecology* 1:5-18.

Franklin, J. F., D. A. Perry, T. D. Schowalter, M. E. Harmon, A. McKee, and T. A. Spies. 1989. Importance of ecological diversity in maintaining long-term site productivity. In *Maintaining the long-term productivity of Pacific Northwest forest ecosystems,* eds. D. A. Perry, R. Meurisse, B. Thomas, R. Miller,

J. Boyle, J. Means, C. R. Perry, and R. F. Powers, chapter 6. Portland, OR: Timber Press.

Frelich, L. E., and C. G. Lorimer. 1991. Natural disturbance regimes in hemlock-hardwood forests of the Upper Great Lakes Region. *Ecological Monographs* 61:145-164.

Frothingham, E. H. 1914. White pine under forest management. *USDA Bulletin* 13.

Gephart, J. S., J. B. Tevik, R. D. Adams, and W. E. Berguson. 1990. *Aspen supply in Minnesota 1977-2007*. Duluth: Natural Resources Research Institute, University of Minnesota.

Graham, S. A., R. P. Harrison, and C. E. Westell, Jr. 1963. *Aspens: Phoenix trees of the Great Lakes region*. Ann Arbor: University of Michigan Press.

Hamburg, S. T., and C. V. Cogbill. 1988. Historical decline of red spruce populations and climatic warming. *Nature* 331:428-431.

Hare, F. K., and M. K. Thomas. 1979. *Climate Canada*. Toronto: John Wiley & Sons Ltd.

Harris, L. D. 1984. *The fragmented forest*. Chicago: University of Chicago Press.

Heinselman, M. L. 1973. Fire in the virgin forests of the boundary waters canoe area, Minnesota. *Quaternary Research* 3:329-382.

Kutzbach, J. E. 1987. Model simulations of the climatic patterns during the deglaciation of North America. In *North America K-3*, eds. W. F. Ruddiman and H. E. Wright, Jr., 425-446. Decade of North American Geology Series. Boulder, CO: Geological Society of North America.

Little, E. L. 1971. *Atlas of United States trees*. Miscellaneous Publication No. 1146. Washington, DC: USDA Forest Service.

Lorimer, C. G. 1989. Relative effects of small and large disturbances on temperate hardwood forest structure. *Ecology* 70:565-575.

Loucks, O. L. 1970. Evolution of diversity, efficiency, and community stability. *American Zoologist* 10:17-25.

Marquis, D. A. 1978. Application of uneven-aged silviculture and management on public and private lands. In *Timber Management Research*, 25-61. Washington, DC: USDA Forest Service.

Maycock, P. F., and J. T. Curtis. 1960. The phytosociology of boreal conifer-ehardwood forests of the Great Lakes region. *Ecological Monographs* 30:1-35.

McInnes, P. F., R. J. Naiman, J. Pastor, and Y. Cohen. 1992. Effects of moose browsing on vegetation and litterfall of the boreal forest, Isle Royale, Michigan, USA. *Ecology.* 73:2059-2075.

Melillo, J. M., J. D. Aber, and J. F. Muratore. 1982. Nitrogen and lignin control of hardwood leaf litter decomposition dynamics. *Ecology* 63:621–626.

Mitchell, H. L., and R. F. Chandler. 1939. The nitrogen nutrition and growth of certain deciduous trees of northeastern United States. *Black Rock Forest Bulletin* 11.

Mladenoff, D. J. 1987. Dynamics of nitrogen mineralization and nitrification in hemlock and hardwood treefall gaps. *Ecology* 68:1171-1180.

————. 1990. The relationship of the soil seed bank and understory vegetation in old-growth northern hardwood-hemlock treefall gaps. *Canadian Journal of Botany* 68:2714-2721.

Mladenoff, D. J., and G. E. Host. 1993. Ecological applications of remote sensing and GIS for ecosystem management in the northern Lake States. In *Forest ecosystem management at the landscape level: The role of remote sensing and GIS in resource management planning analysis and decision-making*, ed. V. A. Sample. Washington, DC: Island Press. In press.

Mladenoff, D. J., and F. Stearns. 1993. Eastern hemlock regeneration and deer browsing in the northern Great Lakes region: A re-examination and model simulation. *Conservation Biology*. In press.

Mladenoff, D. J., M. W. White, J. Pastor, and T. R. Crow. 1993. Comparing spatial pattern in unaltered old-growth and disturbed forest landscapes for biodiversity design and management. *Ecological Applications* 3:293-305.

Morris, R. F. 1963. The dynamics of epidemic spruce budworm populations. *Memoirs of the Entomological Society of Canada* 31:1–332.

Naiman, R. J., C. A. Johnston, and J. C. Kelley. 1988. Alteration of North American streams by beaver. *BioScience* 38:753-762.

Naiman, R. J., J. M. Melillo, and J. E. Hobbie. 1986. Ecosystem alteration of boreal forest streams by beaver (*Castor canadensis*). *Ecology* 67:1254-1269.

Nichols, G. E. 1935. The hemlock-white pine-northern hardwood region of eastern North America. *Ecology* 16:403-422.

Niemi, G. J., and J. R. Probst. 1990. Wildlife and fire in the upper Midwest. In *Management of dynamic ecosystems*, ed. J. M. Sweeney, 31-46. West Lafayette, IN: The Wildlife Society.

Noss, R. F. 1987. Protecting natural areas in fragmented landscapes. *Natural Areas Journal* 7:2-13.

Noss, R. F., and L. D. Harris. 1986. Nodes, networks and MUMs: Preserving diversity at all scales. *Environmental Management* 10:299-309.

Ohmann, L. F., H. O. Batzer, R. R. Buech, D. C. Lothner, D. A. Perala, A. L. Schipper, Jr., and E. S. Very. 1978. Some harvest options and their consequences for the aspen, birch, and associated conifer forest types of the Lake States. Gen. Tech. Rep. NC-48. St. Paul: USDA Forest Service, North Central Forest Experiment Station.

Oosting, H. J. 1956. *The study of plant communities*. San Francisco: W. H. Freeman & Co.

Pastor, J., J. D. Aber, C. A. McClaugherty, and J. M. Melillo. 1984. Aboveground production and N and P cycling along a nitrogen mineralization gradient on Blackhawk Island, Wisconsin. *Ecology* 65:25-28.

Pastor, J., and M. Broschart. 1990. The spatial pattern of a northern conifer-hardwood landscape. *Landscape Ecology* 4:55-68.

Pastor, J., B. Dewey, R. J. Naiman, P. F. McInnes, and Y. Cohen. 1993. Moose browsing and soil fertility in the boreal forests of Isle Royale National Park. *Ecology* 74(2): 467-480.

Pastor, J., R. H. Gardner, V. H. Dale, and W. M. Post. 1987. Successional changes in nitrogen availability as a potential factor contributing to spruce declines in boreal North America. *Canadian Journal of Forest Research* 17:1394-1400.

Pastor, J., and D. J. Mladenoff. 1992. The southern boreal–northern hardwood border. In *A systems analysis of the global boreal forest*, eds. H. H. Shugart, R. Leemans, and G. B. Bonan, 216-240. Cambridge: Cambridge University Press.

———. 1993. Modeling the effects of timber management on population dynamics, diversity, and ecosystem processes. In *Modeling sustainable forest ecosystems*, ed. D. C. LeMaster. Washington, DC: Society of American Foresters.

Pastor, J., and R. J. Naiman. 1992. Selective foraging and ecosystem processes in boreal forests. *American Naturalist* 139:690-705.

Pastor, J., R. J. Naiman, B. Dewey, and P. F. McInnes. 1988. Moose, microbes and the boreal forest. *BioScience* 38:770-777.

Pastor, J., and W. M. Post. 1986. Influence of climate, soil moisture and succession on forest carbon and nitrogen cycles. *Biogeochemistry* 2:3-28.

———. 1988. Response of northern forests to CO_2-induced climate change. *Nature* 334:55-58.

Perala, D. A. 1977. Manager's handbook for aspen in the North Central States. Gen. Tech. Rep. NC-36. St. Paul: USDA Forest Service, North Central Forest Experiment Station.

Raup, H. M. 1941. Botanical problems in boreal North America. *Botanical Review* 7:147-248.

Regier, H. A., and G. L. Baskerville. 1986. Sustainable development of regional ecosystems degraded by exploitive development. In *Sustainable development of the biosphere*, eds. W. C. Clark and R. E. Munn, chapter 3. International Institute for Applied Systems Analysis, Laxenburg, Austria. Cambridge: Cambridge University Press.

Rowe, J. S. 1972. Forest regions of Canada. Canadian Forestry Service publication 1300. Ottawa: Canadian Department of Environment.

Runkle, J. R. 1982. Patterns of disturbance in some old-growth mesic forests of eastern North America. *Ecology* 62:1041-1546.

Shugart, H. H., T. R. Crow, and J. M. Hett. 1973. Forest succession models: A rationale and methodology for modeling forest succession over large regions. *Forest Science* 19:203-212.

Smith, D. M. 1962. *The practice of silviculture.* New York: John Wiley & Sons.

Solomon, A. M. 1986. Transient response of forest to CO_2-induced climate change: Simulation modeling experiments in eastern North America. *Oecologia* 68:567-579.

Sprugel, D. G. 1976. Dynamic structure of wave-generated *Abies balsamea* forests in the northeastern United States. *Ecology* 64:889-911.

Stearns, F. 1949. Ninety years change in a northern hardwood forest in Wisconsin. *Ecology* 30:350-358.

————. 1986. Ecological view of the second hardwood forest and implications for the future. In *Proceedings, Conference on the Northern Hardwood Resource: Management and Potential*, 51-66. Houghton: Michigan Technological University.

————. 1990. Forest history and management in the northern Midwest. In *Management of dynamic ecosystems*, ed. J. M. Sweeney, 107-122. West Lafayette, IN: The Wildlife Society.

Stewart, H., and D. Swan. 1970. Relationships between nutrient supply, growth, and nutrient concentrations in the foliage of black spruce and jack pine. Woodlands Papers no. 19. Pointe Claire, Quebec: Pulp and Paper Research Institute of Canada.

Stoeckeler, J. H., R. O. Strothmann, and L. W. Krefting. 1957. Effect of deer browsing on reproduction in the northern hardwood-hemlock type in northeastern Wisconsin. *Journal of Wildlife Management* 21:75-80.

Temple, S. A., and J. R. Cary. 1988. Modeling dynamics of habitat-interior bird populations in fragmented landscapes. *Conservation Biology* 2:340-347.

Tubbs, C. H. 1977. Natural regeneration of northern hardwoods in the northern Great Lakes region. Research paper NC-150. St. Paul: USDA Forest Service, North Central Forest Experiment Station.

Tubbs, C. H., and F. T. Metzger. 1969. Regeneration of northern hardwoods under shelterwood cutting. *Forestry Chronicle* 45:333-337.

U.S. Department of Commerce. 1968. *Climatic atlas of the United States.* Washington, DC: Superintendent of Documents.

Van Cleve, K., and J. Zasada. 1976. Response of 70-year-old white spruce to thinning and fertilization in interior Alaska. *Canadian Journal of Forest Research* 6:145-152.

Walter, H. 1979. *Vegetation of the earth.* New York: Springer-Verlag.

Webb, T., III. 1974. A vegetational history from northern Wisconsin: Evidence from modern and fossil pollen. *American Midland Naturalist* 92:12-34.

Weetman, G. F. 1968. The nitrogen fertilization of three black spruce stands. Woodlands Papers no. 6. Pointe Claire, Quebec: Pulp and Paper Research Institute of Canada.

Wein, R. W., and M. A. El-Bayoumi. 1982. Limitations to predictability of plant succession in northern ecosystems. In *Resources and dynamics of the bor-*

eal zone, eds. R. W. Wein, R. R. Riewe, and I. R. Methven, 214-225. Methven, Ottawa, Ontario: Association of Canadian Universities for Northern Studies.

Williams, M. 1989. *Americans and their forests: A historical geography.* Cambridge: Cambridge University Press.

6

Landscape Ecosystem Classification: The First Step toward Ecosystem Management in the Southeastern United States

Steven M. Jones
Department of Forest Resources, Clemson University

F. Thomas Lloyd
USDA Forest Service, Southeastern Forest Experiment Station

In the southeastern United States, public demands from forest ecosystems are expanding to include values increasingly associated with complex, diverse forests. Management for this broadened array of values will require a new approach to forest management in the South: ecosystem management. Ecosystem management is a rationale to forest management and silvicultural practice that is guided by the concept of the ecosystem. Tansley (1935) said that the forest ecosystem includes the organism complex of all biota (not just the trees) and the physical environment that supports it. The term *ecosystem management* does not mean management of all organisms in these communities or the edaphic attributes that support them, but rather the consideration of the effects on these communities and environments of a limited set of actions designed to organize the state of the ecosystem in a specific way. In this context, an ecosystem is unbounded by scale (Kimmins 1987), which means that ecosystem management is operational at the stand, landscape, and regional scale, simultaneously. This is a complex concept to incorporate into forest management thinking and planning.

One important obstacle to the adoption of an ecosystem approach is a lack of understanding of the potential of the land to support various forest communities. The extensive alteration of vegetation in the South obscures our view of this potential. Before this complex task of ecosystem management can be considered, we must develop a system of land classification that reflects the potential of the land to support natural biotic communities. In this chapter, we introduce landscape ecosystem classification, a new method of land classification for southern forested ecosystems that goes beyond the traditional use of site index to reflect regional climate, landform, and soil considerations to segment the landscape broadly into units for which more refined interpretations can be made. Through the broad application of this approach, a biologically based context for ecosystem management can be developed.

CHANGING VALUES

The move toward ecosystem management in the South is being driven by a growing realization among the public of the many functions provided by forests. An excellent example of the recognition of nontraditional values is the public's willingness to pay for hunting opportunities. Across the South, the average hunting lease on private lands is $2.16 per acre (Stucky and Guynn 1992). An economic analysis of two counties in South Carolina revealed that expenditures for hunting on private lands contributed substantially to the economy of these counties in the 1990-91 hunting season. In fact, hunter expenditures in both counties exceeded the 1990 total cash receipts from agriculture (Richardson, Yarrow, and Smathers 1992). Estimates indicate that nonconsumptive wildlife expenditures in the southeastern states often rival or sometimes even exceed those of hunter expenditures (Fish and Wildlife Service 1988). These data reflect the economic importance of accommodating a nontimber commodity in a natural resource management strategy. Similar examples could be built for other nontimber commodities, such as recreation.

Land ownership has a tremendous effect on the demands placed on the forest ecosystem. The values derived from various ownerships reflect the personal values of the landowner. For example, private, nonindustrial landowners account for 67 percent of the forest land, but these landowners are the least likely to use an ecosystem approach to management. In surveys of four southeastern states, 22 to 40 percent of all forest owners surveyed considered income from timber sales as the primary benefit from their land (Doolittle 1986, 1988, 1990, 1991). In contrast, 16 to 42 percent considered ownership, beauty, and investment as the primary benefits, which do not require active resource management. Through the regular implementation of Best Management Practices (BMPs), generalized guidelines for soil and water conservation, these owners display an attitude of good forest stewardship and a concern for the ecosystem.

An evaluation of this voluntary guideline program in 1990 revealed a BMP compliance level of 82 percent for private, nonindustrial forest landowners in South Carolina (Hook et al. 1991). The report concluded that making professional advice available to landowners is a viable means of improving BMP compliance. Only 27 percent of the landowners with a low compliance rating employed a professional forester in the development of management plans, as compared to 82 percent with a high compliance rating. In addition, the Forest Stewardship Act of 1990, which created the Forest Stewardship and Stewardship Incentives Programs, is intended to encourage a multiple resource approach to management on private nonindustrial forest lands. In 1991, the first full year of the program, the thirteen southern states reported 2,022 forest stewardship plans covering 377,820 acres (D. A. Hoge, U.S. Forest Service, Southern Region, Atlanta, Georgia, letter, March 2, 1992). Although these two programs do not specifically address ecosystem management, they are encouraging sustainable forestry.

Industrial forest landowners in the Southeast represent 23 percent of the land ownership. Although timber commodity is and always will be the major value, it is no longer the only value. Conservation of soil and water resources is a primary consideration on industry lands, as is sustaining productivity. This is exemplified by the fact that the BMP compliance rate has been 95 percent on industry lands (Hook et al. 1991). More importantly, industry foresters are starting to consider the effect of intensive forestry on other values and are attempting to mitigate this impact within the constraint of maintaining an acceptable rate of return for their investors (Taylor and Owen 1991).

Public lands account for only 10 percent of the forest land in the Southeast; however, it is on public lands that ecosystem management approaches have the greatest potential for immediate application. Operating in an economic environment that does not demand a high rate of return, resource managers on public lands can and should be given the opportunity to provide society with noncommodity values. This may require reductions in harvest levels on some ranger districts. Where compatible, it is possible to provide society with timber along with desired noncommodity values. Research on public lands should address the question of the compatibility of managing for noncommodity and timber values on the same landscape within social, economic, and biological contexts. In time, the technology can be transferred as the need grows on industrial and nonindustrial private lands.

Influencing practices on all of these lands is an increasingly widespread public appreciation for the role of ecosystems in providing useful services. The role of forests and forested wetlands in improving air and stream quality and wildlife and fish habitat is well known. Additionally, the public is coming to appreciate the value of lesser-known species, in particular rare or endangered species, and the role of complex ecosystems in maintaining a healthy forest. For example, the southern pine beetle epidemic is testimony, in part, to relatively low diversity at the stand and landscape level. Although low diversity in the southern pine ecosystem may not be the reason for the epidemic, increasing species diversity within a stand or increasing landscape-level diversity by favoring hardwood forests on those sites that can support commercially productive hardwoods serves to slow down or even prevent the significant spread of infection of the southern pine beetle (Belanger and Malac 1980; Lorio 1980). The restoration of a complex, healthy forest ecosystem in the South will require a new way of looking at the landscape that recognizes ecological heterogeneity and seeks to use that variability in a management context.

THE NEED FOR ECOLOGICAL LAND CLASSIFICATION

The ability to identify the capability and limitations of a given unit of land is still a major limitation to southern forest management on all ownerships. Site index predictions are frequently, and unbeknownst to the user, associated with a high degree of error. Individuals in a hardwood forest that has been repeatedly highgraded are not reliable indicators of land capability. Mismatch of species to site results in economic losses, particularly in areas such as the upper coastal plain, where landowners are encouraged to plant loblolly pine on soils suited to longleaf pine.

The long history of exploitative land use, highly disturbed natural forests, and extensive acreage of plantation forests has obscured relationships between soil and landform characteristics and vegetation. Since the early soil-site studies of Coile in 1935 and Turner in 1938, the need to identify the productivity of southern forest lands has been well understood. Unfortunately, the subsequent decades of research in the South on single- and multiple-factor approaches have resulted in limited success in expressing and predicting forest productivity from soil and site factors (Stone 1984). An excellent history of forest site classification is given by Van Lear (1991).

Emerging from sixty years of research and experience is an "intuition" about the influence of soil and site factors on vegetation productivity, but no widely accepted management tool has been developed to express these relationships. Foresters in the Southeast have developed skeptical attitudes toward the potential application of ecological land classification, even given the acceptance and success in other regions of the United States. As a result, site index continues to be the method of choice in spite of its shortcomings and esoteric qualities.

Integrated ecological land classification serves as a management tool allowing resource managers to identify the potential of a given unit of land for a wide range of values. The concept of ecosystem management is not totally embodied in ecological land classification. However, such an approach is a necessary precursor to management practices that are designed to enhance or feature nontimber values. An approach to ecological land classification that is showing promise in the South is landscape ecosystem classification.

THE LANDSCAPE ECOSYSTEM
CLASSIFICATION APPROACH

Landscape ecosystem classification (LEC) was initiated in Michigan in the mid-1970s (Barnes et al. 1982). In South Carolina, as with the habitat-type approach in the western United States (see chapter 8), early efforts in the mid-1970s were driven by vegetation. By the late 1970s, however, LEC had evolved into an integrated approach. The Michigan efforts developed the terminology of *landscape ecosystem classification*, which has been adopted for modeling applications in the Southeast. LEC expresses the interrelationships (1) between vegetation and landform, (2) between vegetation and soils, and (3) between landform and soils. The term *landscape* is used as a modifier to emphasize that ecosystems are geographic units extending horizontally over the land (Barnes 1989). Landform is the key component because it is permanent and relatively easy to recognize.

Plants can be considered as integrators of environmental factors. In the absence of disturbance, the distribution of individual species in competition with their associates is a function of environmental conditions. Those species with a narrow ecological amplitude are considered "diagnostic" and are indicative of particular environmental conditions. Species with a broad ecological amplitude are considered "constant" species and are not indicative of a certain set of environmental conditions (Mueller-Dombois and Ellenberg 1974). Species with similar environmental requirements have overlapping distributions and form associations. Associations of diagnostic species under undisturbed conditions are considered to be "site indicators." The presence and absence of diagnostic species are used in place of timber productivity (site index) as a means of determining which land units are equivalent in terms of potential biological productivity.

Under relatively undisturbed conditions, associations of diagnostic species (vegetation types) are related to landform and soils. Landform factors may include slope gradient, slope position, aspect, and slope shape, and the soil component may include drainage, chemistry, and physical properties, such as depth of clay, amount of clay, or thickness of sandy epipedon. Because the interrelationships of vegetation, landform, and soil are known, the resulting land classification is ecologically based. Once the model is developed, it can be applied in the absence of forest vegetation or in disturbed or plantation forests with the permanent features of landform and soil.

This approach to ecological land classification takes into account variation due to major environmental factors by recognizing physiographic provinces, regions, and subregions. For instance, within a given physiographic region or subregion, major climatic patterns would not significantly vary. Likewise, when parent material differences are known to affect major soil properties and alter plant species composition and productivity, lands are subdivided into physiographic regions or subregions. Since the vegetation, soil, and landform relationships vary by region, the discriminating soil and landform variables may differ by region. Within a region, climate and dominant parent material generally do not vary. Occasionally, however, regions with particularly complex climatic, geological, or other patterns must be subdivided into physiographic subregions and the regional LECs must be refined to reflect the subregional differences. The physiographic classification project throughout the South initiated in 1973 by the Southern Forest Environmental Research Council (Hodgkins, Golden, and Miller 1976; Pehl and Brim 1985; Myers, Zahner, and Jones 1986; Miller and Golden 1991) is a hierarchical accounting of the major factors of climate and parent material in the southeastern United States.

The LEC approach is hierarchical and can be structured to accommodate the proposed National Hierarchical Framework of Ecological Units being developed by the USDA Forest Service (W. H. McNab, USDA Forest Service, Bent Creek Experimental Forest, Asheville, North Carolina, review draft, October 5, 1992). Regions and subregions are further classified into landtype associations. The landtype association expresses differences in parent material, major soil-forming processes, and relief (e.g., bottomlands and uplands). Within each landtype association, finer divisions can be made; these are landtypes. An example of landtype would be classifying uplands into two broad groups of xeric and mesic. Each landtype is subdivided into landtype phase. In LEC terminology, the landtype phases are recognized as site units or landscape ecosystem units. Landtype phases (site units) are identified on the basis of soil physical properties or micro- and meso-level landform properties, such as aspect, slope position, slope gradient, or slope shape. The site, which is known traditionally as the site type, is the lowest level in the hierarchy. Sites or site types can be morphologically different but similar in productivity because of the interaction of soil and landform variables. For instance, protected, north-facing slopes with thin soils may be similar in productivity with exposed, south-facing slopes with thick soils. These two site types are similar in productivity and are capable of supporting the same vegetative association; therefore, they are classified into the same landtype phase (site unit). As a result, within a given LEC model there may be dozens of possible sites (site types) that are grouped into four, five, or six landtype phases (site units). This approach serves to simplify the landscape, and the landtype phase (site unit) is the level where individual stand management considerations are made.

THE DEVELOPMENTAL PROTOCOL

The development and implementation of an LEC is a four-phase procedure.

- Phase I: Identification of the site units from relatively undisturbed vegetation and soil and identification of discriminating landform and soil variables
- Phase II: Identification and description of the various successional vegetation types for each site unit
- Phase III: Mapping of site units on the ground and through Geographic Information Systems applications
- Phase IV: Development of management interpretations for each site unit

In developing an LEC, it is necessary to proceed sequentially through phases I and II; however, phases III and IV may occur concurrently.

An LEC model has been developed on the Savannah River Site in South Carolina (Jones, Van Lear, and Cox 1981, 1984; Jones and Van Lear 1984; Van Lear and Jones 1987) and expanded to include the Hilly Coastal Plain Province (Jones 1990). Phases I and II have been completed, although some refinement is in progress. Phases III and IV are in progress, and recent work has been conducted to develop implementation of the model through geographic information systems (GISs) (Lloyd, Chubb, and Jones 1990; Jones 1991). Within the uplands of the Hilly Coastal Plain, soils are the driving variable in the model, and landform influences are most apparent on slopes adjacent to stream drainages.

The landscape ecosystem approach has also been successfully applied through phases I and II within the Piedmont of South Carolina (Gay 1992; Jones 1988, 1991) and is being applied within the Georgia Piedmont. Current research is under way to test the hypothesis that LEC models can be incorporated into Piedmont growth and yield models (Lloyd 1991) for mixed stands of hardwoods and pines (phase IV). The Piedmont model is a function of a combination of landform and soil factors, and landtype associations are subdivided into landtypes and landtype phases (site units) on the basis of interacting soil and landform characteristics.

Within the southern Appalachians, phase I efforts of model development have been completed on the Nantahala National Forest, Highlands Ranger District in North Carolina (Gattis 1992), and are under way on the Chattahoochee National Forest, Chattooga and Blairsville Ranger Districts in Georgia. Modeling efforts are also under way in the southern

Appalachians on the Pisgah National Forest and Bent Creek Experimental Forest, North Carolina, by McNab of the Southeastern Forest Experiment Station (McNab 1991). Results indicate that landform (elevation and protection) is the major discriminator of site differences but that soils (solum thickness) also account for significant variability.

The LEC approach is currently being tested by Steven Jones within forested wetland ecosystems within the South Carolina coastal plain. Within the upper Edisto River basin, landtype phases (site units) seem to be strongly related to soil taxonomy, internal drainage characteristics, and the presence of organic horizons. Similar relationships are being identified within the Coastal Flatwoods Region on Union Camp Corporation lands in Jasper and Beaufort Counties.

RESEARCH APPLICATIONS

The site units identified in the LEC models are based on measurable structural attributes of the landform, soil, and vegetation. As various research efforts utilize the LEC model as a framework in experimental designs, similarities and differences among the site units in functional attributes are revealed. This knowledge improves our ability to interpret and predict the consequences of management actions.

In the examination of root distribution of loblolly pine in South Carolina, McCollum (1992) revealed a significant relationship with site units. Intermediate site units had fewer roots (76 roots per square meter) than did xeric site units, which had fewer roots (224 roots per square meter) than did subxeric site units (314 roots per square meter). This functional relationship was explained by considering the environmental gradient as an available soil water gradient. Individual trees on intermediate sites have a greater supply of soil water available for growth; as a result, fewer roots are needed to supply the aboveground portion of the tree with adequate water. The individual tree partitions a greater proportion of biomass into aboveground growth; thus, trees on intermediate sites are taller. On xeric and subxeric site units, soil water deficits are common, which requires a greater production of root biomass to provide the aboveground stem with water for growth. Xeric sites experience such high soil water tensions that root mortality is high, resulting in fewer roots on the xeric site unit than on the subxeric site unit.

A dendroclimatic examination of white oak (*Quercus alba*) in relatively undisturbed Piedmont hardwood stands was conducted across four site units (Jacobi and Tainter 1988). Ring width chronologies of individuals growing on xeric and subxeric site units displayed greater climatic stress than did chronologies of individuals on intermediate and submesic site units.

A recent study in Piedmont hardwood stands described the forest floor and humus form across the xeric, subxeric, and intermediate site units (Baldwin-Ball 1992). The humus form taxonomy was developed by Klinka et al. (1981) in British Columbia. On xeric and subxeric site units, the humus form was identified as mormoder; on intermediate sites the humus was a mull.

These research applications serve to substantiate the functional uniqueness of site units based on structural characteristics. In addition, when research is designed in the framework of LEC, it is possible for researchers to identify where on the landscape their research results can be successfully applied and to compare results among studies.

THE LEC AS A TOOL IN DECISION MAKING

An important product of the planning process is a desired future condition. The future state may be an economic, a social, or a biological condition that is desired by those involved with or affected by the planning process. To identify a desired future condition, decision makers need a rational approach to deal with encountered complexities. The prediction and subsequent definition of a desired future condition can be facilitated through LEC.

The applications of LEC are revealed by first identifying the decision makers. Second, it is necessary to identify the social or political scale under which the institutions of the decision makers (private individuals, nonforest industry, forest industry, and the public) are operational. The various institutions function at regional, state, and local scales (table 6.1).

The desired future condition on private (individual and nonforest industry) lands is driven by a multitude of factors depending on the needs and desires of the landowners. As previously documented, there are private landowners who have an interest in sustainable forest management, but they are few. At the local level, the application of LEC can occur only through the actions of private forestry consultants. These private consultants are required in many southern states to maintain their professional credentials as registered foresters through continuing education credit.

TABLE 6.1. **Decision Makers for Forest Land Recources According to Organizational Affiliation and Political Scale**

	Organizational Unit		
Scale	*Private*	*Industry*	*Public*
Regional	Federal agency	Corporate office	Regional office
State	State & regulatory planning board	Timberland division	Supervisor's office
Local	County & private citizen	Forest manager	District silviculturist

The responsibility of technology transfer to the practicing professional lies largely with University Cooperative Extension. These continuing education programs need to broaden their view to include or emphasize programs on ecosystem management and forest sustainability.

In the Southeast, it is not likely that sustainable forestry or ecosystem management is going to be mandated through legislative action. However, sustainable development considerations at the regional and state level are influencing the situation at the local level. Most notably at the regional level, protection of forest wetlands through the Clean Water Act has greatly affected the development and use of private forest lands. In fact, in the South it is the private landowner who has been given the responsibility of providing society with clean water and other wetland benefits because the large majority of land ownership is private.

On forest industry lands there is great potential for LEC, particularly at the local or forest manager level. Of course, the biggest application potential is in improving predictions of site quality and growth and yield predictions for commercially important timber species. Commercially important species include both pines and hardwoods because in recent years hardwoods have been used in the pulp process. There may also be potential for predicting hazard ratings for insects and disease, particularly the southern pine beetle. Most prediction equations used in the South include various soil and landform features or the site index along with stand structural conditions in an effort to forecast physiological stress (Belanger and Malac 1980; Lorio 1980). Substituting LEC site units for these various site and soil variables could improve the prediction process.

A secondary benefit of LEC is that, once it is in place, interpretations related to other noncommodity values that are correlated with spatial and temporal variability in the landscape can be made. Progressive companies realize that minor adjustments in accepted management practices on industry forest lands can provide society with other benefits at little or no cost (Hughes 1992; Taylor and Owen 1991) but yield great returns in terms of public relations.

The LEC approach is already in practice on public lands in Georgia and North Carolina. As a result, we have a better appreciation for application potential and how the approach helps to promote an ecosystem awareness in the management of forest resources. We can identify the decision makers at three organizational unit scales, the regional office, the supervisor's office, and the district silviculturist. Decision makers within these organizational units influence determination of the desired future condition at three biological scales, macro/regional, meso/landscape, and micro/site. The regional scale could range in size from a state or perhaps several states (southern Appalachians) to a physiographic region or subregion within a state. The landscape scale could range from a ranger district to a single watershed.

Most decisions at the regional office level are directed at determining the desired future condition at the regional and landscape scales. At the supervisor's office level, decisions aimed at setting the desired future condition should take into account pertinent available information at the regional, landscape, and site levels. The district silviculturist working at the site level should take into account both site-level information and landscape-level information. This decision-making process relies heavily on maps, which serve to synthesize information and facilitate communication. In fact, the major criterion in measuring the usefulness of a classification approach is its adaptability to mapping procedures and the production of accurate, useful maps from which interpretations of land productivity and other resource values can be made.

Obviously, mapping site units based on potential climax vegetation would have limited use in the identification of land productivity in the southern United States. As a result of intensive forestry, including widespread conversion to pine forests, and other widespread anthropogenic effects, the South's forests are predominantly composed of successional species whose presence is a reflection of the interaction of disturbance with environmental conditions.

The singular use of soil survey is also perceived as having limited application in delineating site units with similar productive potential. Soil taxonomy is often criticized because soil series are classified based on morphological features often unrelated to site productivity. This problem is overcome, in part, by combining soils at the series level into groups that represent ecological equivalents, that is, those soils that produce the same type of late-successional vegetation on a given landform.

Early results of efforts to integrate LEC models into geographic information system data layers are promising (Lloyd, Chubb, and Jones 1990). Landform is expressed as digital elevation data, and soils are expressed through digitized soil survey data that are grouped and remapped according to their ecologically equivalent groups. Predicted site unit boundaries and potential vegetation are mapped through modeling the interaction of landform and soils. Mapped site unit boundaries can be refined in the field by observing the distribution of diagnostic species when they have not been eliminated through land-use practices. Since LEC simplifies all of the various combinations of soil variables and landform variables into relatively few site units for a given region, interpreting mapped landscapes is less confusing.

These site units can be used to identify potential regional- and landscape-level information that may be necessary, in addition to site-level information, to set the desired future condition for a forest. For example, the USDA Forest Service, Savannah River Station, is responsible for management of the forest resources of the 200,000-acre Savannah River Site, South Carolina. The Savannah River Site lies within the Hilly Coastal Plain Province, Upper Loam Hills Region.

In setting the desired future condition for the Savannah River Site, the Savannah River Station must consider biological diversity issues. At the landscape level, biological diversity encompasses both spatial and temporal aspects. For example, early seral, midseral, and old-growth stands can occur on xeric, intermediate, and mesic sites. Thus, a landscape approach to management must go beyond the stand or site level to consider these spatiotemporal aspects of diversity. Biodiversity-related problems are not going to be solved solely by attempting to enhance species richness at the stand level.

To illustrate the utility of LEC in addressing landscape-level biodiversity issues at the Savannah River Site, let us consider a stand occurring on the mesic site unit (figure 6.1). If the decision maker were to set the desired future condition of this stand based on stand-level (micro-scale) information only, the chosen alternative would most likely be loblolly pine culture. This decision would be driven by the productive capacity of the soils on the mesic site unit and the economic value of pine.

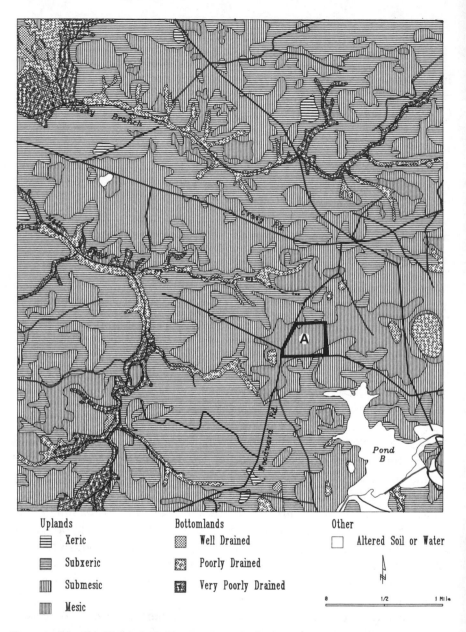

Uplands		Bottomlands		Other	
☰	Xeric	▨	Well Drained	☐	Altered Soil or Water
☰	Subxeric	▨	Poorly Drained		
▥	Submesic	▨	Very Poorly Drained		
▨	Mesic				

FIGURE 6.1. **Predicated Site Units (Landtype Phases) Map of the Mill Creek Area, Savannah River Site, South Carolina.**
The stand example described in the text is identified by the letter A.

Productive soils in the Hilly Coastal Plain have been in agriculture for many decades. These soils are well drained and have a thin sandy surface layer (less than 20 inches) over a sandy clay loam subsurface. Agriculture on these soils is the predominant land use in both the Hilly Coastal Plain and the Middle Coastal Plain Provinces. Because of land-use characteristics on a regional scale, late-successional hardwood stands on the mesic site unit occur rarely.

Examination at the landscape scale, the Savannah River Site, reveals that the mesic site unit on upland flats is a relatively low percentage of the total acreage. Additionally, in the early 1950s, when the Savannah River Site came under federal ownership, the former agriculture areas were planted in southern pines. As a result, the present stand condition for the mesic site unit on upland flats is loblolly pine. When regional- and landscape-scale information is integrated into the decision-making process, it becomes clear that, in terms of promoting biological diversity, a high priority should be given to a hardwood stand as the desired future condition on the site in question.

We must recognize that promoting biological diversity to this degree is realistic only on public lands. Economics will probably drive the desired future condition on forest industry lands to loblolly pine plantation and on private lands to row crops. An exception to pine plantation as the desired future condition on forest industry lands is seen within the Hilly Coastal Plain for Union Camp Corporation lands. Their pulp mill at Eastover, South Carolina, is utilizing more than 80 percent hardwoods in the paper-making process; therefore, early successional hardwoods, primarily sweetgum (*Liquidambar styraciflua*), is an economically realistic desired future condition. Once again, LEC can be used to identify these productive, upland site units within the market area of the Eastover Mill.

AN ECOSYSTEM APPROACH TO MANAGEMENT IN THE SOUTHEASTERN UNITED STATES

We have documented the need for an ecosystem approach to forest resources management and have demonstrated the potential for application in the southeastern United States. We must bear in mind that our expectations should vary depending on land ownership. On private lands, expectations should be to maintain ecosystem integrity. This implies maintaining the productive capacity of the land through soil conservation measures and maintaining clean water and air. On forest industry lands, we should expect maintenance of ecosystem integrity and recognition of the influence of certain forest practices on noncommodity values. On public lands our expectations should include providing other noncommodity benefits along with recognizing the effect of forest practices on noncommodity values and the maintenance of ecosystem integrity.

Landowners, foresters, and society should recognize that not all values can be produced on a single forest stand, landscape, or, for some values, even a collection of contiguous landscapes, such as a ranger district. As a result, in the decision process of determining a desired future condition, it is necessary to consider scale, both spatial and temporal. Depending on the scale of the organizational unit, the desired future condition may be for a stand, a landscape, or even a region.

The array of alternatives for desired future conditions across a spatial and temporal gradient can be effectively identified for a given unit of land through LEC. The LEC approach can be applied for any value that varies in conjunction with spatial or temporal variability; therefore, LEC has a wide range of applications. Since the approach has been successfully demonstrated in the coastal plain, piedmont, and mountains, it is time to recognize that single-component approaches, such as commercial timber types, soil types, or descriptive vegetation inventories, cannot satisfy informational needs for ecosystem management at the level of an integrated approach.

If our expectations with respect to ecosystem management and noncommodity values are to be met, incentives and opportunities for landowners and decision makers are necessary. On public lands, this may require a decrease in allowable sale quantity for some districts, which will give district rangers the opportunity to try alternative management practices. On forest industry lands this will require managers at all levels to reevaluate intensive pine culture over vast acreage with no regard to noncommodity values. Innovative companies are considering the effect of management practices on other societal goals of interest and even providing certain benefits while providing an acceptable level of profit margin for their shareholders. To maintain ecosystem integrity on private lands, professional forestry consultants, as well as loggers, must be educated through continuing education sponsored by university extension programs. Economic incentives must be available for private landowners to provide nontimber values. For instance, private landowners are providing wildlife values because of the market-driven demand for hunting leases.

On the other hand, private landowners are providing wetland values through a politically driven process that considers the desires of an urban society at the expense of rural, private landowners.

Finally, we must recognize that we are in a different era than the early post–World War II period. The economic and social conditions in the United States after World War II changed the role of forests, forestry, and foresters. Before World War II, the emphasis was on renewing southern forests in response to a previous period of timber mining and the decline of forest acreage and volume. After World War II, the forests were viewed as a source of industrial wood, supplying the needs of a country in an economic boom involving renovation and reconstruction. Forestry was production oriented, and intensive tree farming was viewed as necessary to save the United States from a predicted timber famine. Plantation forestry became common practice, particularly in the South, where a combination of low land prices, favorable terrain, suitable species that grew rapidly on formerly agricultural land, and well-developed infrastructure generated acceptable financial returns (Sedjo 1991). Forestry schools were producing foresters with the management tools to meet society's demands on the forests for timber and wood products.

Today's social and economic climate is much different from that of the early postwar period, and the role of forests, forestry, and foresters should reflect this difference. The role of the southern forest, particularly its public lands, is being reevaluated by society. Society views the southern forest as providing noncommodity values in addition to the timber commodity. Consequently, not all landowners desire the same future condition for their forest lands. This is requiring a fresh look at forestry practices with modification of some practices and development of new technology to meet the objectives of landowners and the needs of society. This does not imply that the old practices were wrong, but it does imply that new management philosophies and technologies are in order. Today's and tomorrow's forester needs the tools and understanding to identify the array of alternatives in desired future condition and to implement an ecosystem approach to sustainable forest management.

REFERENCES

Baldwin-Ball, R. E. 1992. Distribution and development of root mat and humus form in South Carolina Piedmont forests. Thesis, Department of Forest Resources, Clemson University, Clemson, South Carolina.

Barnes, B. V. 1989. Old-growth forests of the northern lake states: A landscape ecosystem perspective. *Natural Areas Journal* 9(1):45-57.

Barnes, B. V., K. S. Pregitzer, T. A. Spies, and V. H. Spooner. 1982. Ecological forest site classification. *Journal of Forestry* 80:493-498.

Belanger, R. P. and B. F. Malac. 1980. Silviculture can reduce the losses from the southern pine beetle. USDA Agric. Handbk. 576.

Coile, T. S. 1935. Relation of site index for shortleaf pine to certain physical properties of the soil. *Journal of Forestry* 33:726-730.

Doolittle, L. 1986. Nonindustrial private forest owners and resources in east Texas. Unpublished report, Social Science Research Center, Mississippi State University.

_____. 1988. Nonindustrial private forest owners and resources in Mississippi. Unpublished report, Social Science Research Center, Mississippi State University.

_____. 1990. Nonindustrial private forest owners and resources in Tennessee. Unpublished report, Social Science Research Center, Mississippi State University.

_____. 1991. Nonindustrial private forest owners and resources in Alabama. Unpublished report, Social Science Research Center, Mississippi State University.

Fish and Wildlife Service. 1988. 1985 national survey of fishing, hunting, and wildlife associated recreation. US Department of the Interior.

Gattis, J. T. 1992. Landscape ecosystem classification on the Highlands Ranger District, Nantahala National Forest in North Carolina. Thesis, Department of Forest Resources, Clemson University, Clemson, South Carolina.

Gay, J. U. 1992. Landscape ecosystem classification of forested ecosystems within the Carolina Slate Belt Region of South Carolina. Thesis, Department of Forest Resources, Clemson University, Clemson, South Carolina.

Hodgkins, E. J., M. S. Golden, and W. F. Miller. 1976. Forest habitat regions and types on a photomorphic-physiographic basis: A guide to forest site classification in Alabama-Mississippi. *Alabama Agricultural Experiment Station and Mississippi Agricultural and Forestry Experiment Station Southern Cooperative Bulletin 210.*

Hook, D. D., W. McKee, T. Williams, B. Baker, L. Lundquist, R. Martin, and J. Mills. 1991. A survey of voluntary compliance of forestry BMPs. Columbia: South Carolina Forestry Commission.

Hughes, J. 1992. Roles of industry: Balancing environmental concerns and timber supplies. Presented at the Annual Meeting of the Appalachian Society of American Foresters, Asheville, North Carolina, February 12-14, 1992.

Jacobi, J. C., and F. T. Tainter. 1988. Dendroclimatic examination of white oak along an environmental gradient in the Piedmont of South Carolina. *Castanea* 53(4):252-262.

Jones, S. M. 1988. Old-growth forests within the Piedmont of South Carolina. *Natural Areas Journal* 8(1):31-37.

————. 1990. Application of landscape ecosystem classification within the southeastern United States. In *Proceedings of 1989 Society of American Foresters National Convention, Spokane, Washington,* 79-83.

————. 1991. Landscape ecosystem classification for South Carolina. In *Ecological land classification: Applications to identify the productive potential of southern forests,* eds. D. L. Mengel and D. T. Tew, 59-68. Gen. Tech. Rep. SE-68. Charlotte, NC: USDA Forest Service, Southeastern Forest Experiment Station.

Jones, S. M. and D. H. Van Lear. 1984. A habitat type approach for classifying sites within the upper coastal plain of South Carolina. In *Proceedings: Forest land classification: Experience, problems, perspectives,* ed. J. G. Bockheim, 241-250. NCR-102. Madison, WI: North Central Forest Soils Committee.

Jones, S. M., D. H. Van Lear, and S. K. Cox. 1981. Major forest community types of the Savannah River Plant: A field guide. Savannah River Plant, National Environmental Research Park Program, US Department of Energy. SRO-NERP-9.

————. 1984. A vegetation-landform classification of forest sites within the upper Coastal Plain of South Carolina. *Bulletin of the Torrey Botanical Club* 111(3):349-360.

Kimmins, J. P. 1987. *Forest ecology.* New York: Macmillan.

Klinka, K., R. N. Green, R. L. Trowbridge, and L. E. Lowe. 1981. Taxonomic classification of humus forms in ecosystems of British Columbia. Rep. 8. Victoria, BC: British Columbia Ministry of Forest Land Management.

Lloyd, F. T. 1991. Forecasting growth of pine-hardwood mixtures from their ecological land class. In *Ecological land classification: Applications to identify the productive potential of southern forests,* eds. D. L. Mengel and D. T. Tew, 93-95. Gen. Tech. Rep. SE-68. Charlotte, NC: USDA Forest Service, Southeastern Forest Experiment Station.

Lloyd, F. T., R. M. Chubb, and S. M. Jones. 1990. Using GIS-implemented ecological land classification model to determine where pine-hardwood regeneration practices work on the Savannah River Site. Poster presented at the Southern Appalachian Man and the Biosphere Conference, Gatlinburg, Tennessee, November 12-13, 1990.

Lorio, P. L., Jr. 1980. Rating stands for susceptibility to southern pine beetle. In *The southern pine beetle*, eds. R. C. Thatcher, J. L. Searcy, J. E. Coster, and G. D. Hertel, 153-163. Tech. Bull. 1631. Washington, DC: USDA Forest Service.

McCollum, M. M. 1992. Density and distribution of loblolly pine roots along an environmental gradient. Thesis, Department of Forest Resources, Clemson University, Clemson, SC.

McNab, W. H. 1991. Land classification in the Blue Ridge Province: State-of-the-science report. In *Ecological land classification: Applications to identify the productive potential of southern forests*, eds. D. L. Mengel and D. T. Tew, 37-47. Gen. Tech. Rep. SE-68. Charlotte, NC: USDA Forest Service, Southeastern Forest Experiment Station.

Miller, W. F., and M. S. Golden. 1991. Forest habitat regions: Integrating physiography and remote sensing for forest site classification. In *Ecological land classification: Applications to identify the productive potential of southern forests*, eds. D. L. Mengel and D. T. Tew, 73-80. Gen. Tech. Rep. SE-68. Charlotte, NC: USDA Forest Service, Southeastern Forest Experiment Station.

Mueller-Dombois, D., and H. Ellenberg. 1974. *Aims and methods of vegetation ecology.* New York: John Wiley & Sons.

Myers, R. K., R. Zahner, and S. M. Jones. 1986. Forest habitat regions of South Carolina. Clemson University, Department of Forestry, Research Series 42. Map supplement, Scale 1:1,000,000.

Pehl, C. E., and R. L. Brim. 1985. Forest habitat regions of Georgia, Landsat 4 imagery. Georgia Agricultural Experiment Station Special Publication 31. December. Map supplement, Scale 1:1,000,000.

Richardson, C. L., G. K. Yarrow, and W. M. Smathers, Jr. 1992. Economic impact of hunting on rural communities: Jasper and Beaufort Counties in South Carolina. The Extension Wildlife Program, Department of Aquaculture, Fisheries, and Wildlife, Clemson University, Clemson, SC.

Sedjo, R. A. 1991. Forest resources: Resilient and serviceable. In *America's renewable resources: Historical trends and current challenges*, eds. K. D. Frederick and R. A. Sedjo, 81-120. Washington, DC: Resources for the Future.

Stone, E. L. 1984. Site quality and site treatment. In *Forest soils and treatment impacts*, ed. E. L. Stone, 41-52. Knoxville: Sixth North American Forest Soils Converence, University of Tennessee.

Stucky, G., and D. C. Guynn. 1992. Industrial hunting lease programs in the southern United States: 1992. Department of Forest Resources, Clemson University, Clemson, SC.

Tansley, A. G. 1935. The use and abuse of vegetational concepts and terms. *Ecology* 16:284-307.

Taylor, D., and C. Owen. 1991. Stewardship responsibilities and the forest products industry. *Journal of Forestry* 89(11):13-16.

Turner, L. M. 1938. Some profile characteristics of the pine-growing soils of the coastal plain region of Arkansas. *Arkansas Agricultural Experiment Station Bulletin* 361.

Van Lear, D. H. 1991. History of forest site classification in the South. In *Ecological land classification: Applications to identify the productive potential of southern forests*, eds. D. L. Mengel and D. T. Tew, 25-36. Gen. Tech. Rep. SE-68. Charlotte, NC: USDA Forest Service, Southeastern Forest Experiment Station.

Van Lear, D. H., and S. M. Jones. 1987. An example of site classification in the southeastern coastal plain based on vegetation and land type. *Southern Journal of Applied Forestry* 11(1):23-28.

7

Limitations on Ecosystem Management in the Central Hardwood Region

George R. Parker

Department of Forestry and Natural Resources, Purdue University

A LARGE, DIVERSE AREA

The Central Hardwood Region includes about 140 million hectares (340 million acres) of the east-central area of the United States (figure 7.1). The region is dominated by temperate deciduous forest that includes a rich mixture of plant and animal species. East-west moisture and north-south temperature gradients influence the distribution of species on a regional scale, while aspect, slope position, soil characteristics, and disturbance control species composition on local scales (Campbell 1987; Ward and Parker 1989; Parker and Weaver 1989). Occasional severe winter temperatures restrict the northern extent of many plant and animal species, and summer drought restricts species from east to west and to a lesser extent north to south.

Soils of the southern hilly, unglaciated region vary in depth, fertility, and texture. Species composition and richness vary from the productive mesic sites on lower slopes and on north and east aspects to less productive upper slopes and south and west aspects. Specialized sites such as shallow soils with barrens communities also occur. Soils with limestone parent material tend to have greater species richness than do sandstone soils.

Soils of the glaciated northern part of the region are generally deep and fertile. Mesic to wet microgradients result in high species richness on these

FIGURE 7.1. **The Central Hardwood Region.**
After Clark (1989).

glaciated sites. For example, a good-quality 20-hectare woodland will have around two hundred native plant species (Pursell 1989). Prairie communities and wetlands were once common throughout this northern area but have largely been lost to human activities, and remaining forests have been reduced to small woodland fragments or riparian strips along streams.

Most of the forest remaining in the region is capable of producing commercial timber products (Merritt 1980). Much of the region is also suitable for alternative land uses such as urban development and agricultural crop production.

HEAVY HISTORICAL DISTURBANCE

The current structure of forests within this region is largely the result of human activity beginning in the 1500s with Native American activities. Native Americans were active in the region over the last ten thousand years, but written records date only to the 1600s. These people used fire throughout the region to move game animals and to open forest understories (Steyermark 1959; McCord 1970; Ladd 1991; Martin 1989). Fires were set annually, usually in late summer, and must have burned over large areas of the region during dry years.

Native Americans were also agrarian in their use of the landscape. Upland mesic forests were cleared in the northeastern glaciated area and in the river valleys throughout the region for their agricultural activities. Their use of fire and forest clearing maintained prairies and savannas and also changed the species composition of forest communities (Nuzzo 1986). The overstory structure of many of the remaining old-growth remnants, currently dominated by seral oak and hickory species, most likely originated from Native American practices.

The Native American population was relatively small in the region so that human activities, although widespread, left most of the landscape in natural vegetation at any given time. The maximum population of Indiana before European settlement is estimated to have been about twenty thousand individuals. Recent estimates of Native American populations indicate that they may have been ten times greater than the above estimate of twenty thousand (DeVivo 1990). This larger population is more realistic in terms of old-growth forest structure across the region.

European activity starting in the late 1700s initially adopted Native American methods of clearing forests. Much of the land cleared in the early 1800s was abandoned as fertility declined and settlers moved to new areas; it then reverted back to forest. But as technology, such as steam power, developed through the mid-1800s, forests were permanently cleared from the landscape. Overharvesting and fragmentation of the forest resulted in extirpation of many large, wide-ranging mammals (black bear, timber wolf, cougar, elk, bison) from the interior parts of the region by 1860 (Reeves 1976). Abuse of resources and overexploitation was so severe in the 1800s that white-tailed deer were extirpated from Indiana by 1900. The historical record of decline of other plant and animal species during this period is quite limited. Most species were probably able to persist in fragments of habitat across the landscape.

Clearing and widespread resource abuse continued until the late 1930s. Since then, disturbance has been greatly reduced. For example, the area of grazing by domestic livestock has decreased from 70 percent of the land base in the 1930s to around 30 percent today. Annual burning of woodlands, practiced widely by farmers during the first few decades of this century, is now quite limited in the region. Open forest understories, easily seen on aerial photos from the 1930s, are now virtually nonexistent.

Reforestation, particularly in the western and southern parts of the region, has reduced fragmentation of the forest since the 1930s. The incorporation of abandoned farmland into the public domain allowed reforestation of the hilly areas, while land use in relatively flat northern parts of the region has intensified. Little of this area has returned to forest. Currently, about 6 percent of the landscape of central Indiana and Ohio is forested.

Most of the current forests across the region originate from the heavy cutting and agricultural abandonment 60 to 120 years ago and are even aged (uneven size makes them appear uneven aged). Less than 1 percent of the original forest predates European settlement (Parker 1989; Sander, Merritt, and Tyron 1981). Differential growth rates among species after disturbance such as clear-cutting result in diameter distributions similar to those of uneven-aged forests within fifteen to twenty-five years (Leopold, Parker, and Swank 1985).

A HIGHLY FRAGMENTED LANDSCAPE

Although there has been reforestation in the region during the past sixty years, this forest remains fragmented and is still recovering from the widespread abuse from 1800 to the 1930s. Approximately 40 million of the original 140 million hectares are in forest cover today. Human developments such as roads, transmission corridors, and urbanization further fragment the region so that little remote forest land remains (National Outdoor Recreation Supply Information System 1987). For example, the whole eastern half of the United States has only 1.6 million hectares of wilderness and 0.6 million hectares of remote wild lands (lands greater than 3 miles from a road) in federal ownership. State wilderness or remote backcountry in the eastern United States totals about 4 million hectares. Most of these relatively unfragmented lands occur outside the Central Hardwood Region.

The 14,374-hectare landscape shown in figure 7.2 is a Thematic Mapper view of part of the Pleasant Run Unit of the Hoosier National Forest in south-central Indiana. This landscape is part of the largest contiguous forest region of Indiana. Interior forest, at least 200 meters from nonforest edge, represents 27 percent, while forest edge (0 to 200 meters) accounts for 53 percent of the landscape. The nonforest area occurs mostly along streams that are in private ownership and have been cleared for rural homesites and/or agriculture. The small openings scattered through the forest interior vary in their significance depending on the species of interest and the distance from the large, nonforested areas. Although there are a few less-fragmented landscapes in the region, this landscape represents reasonably high-quality forest of the Central Hardwood Region.

The long-term survival of plant and animal species in the current landscape context is not well understood. However, it is likely that simply protecting areas from human activity will not guarantee survival of species in this highly disturbed region. Some form of management such as prescribed fire or control of herbivore (white-tailed deer) populations will be

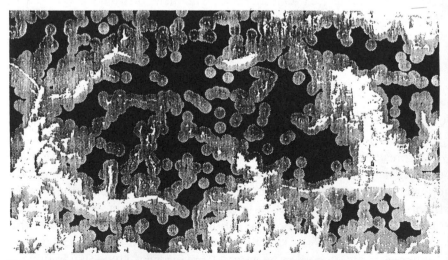

FIGURE 7.2. **A Thematic Mapper View of Part of the Pleasant Run Unit of the Hoosier National Forest in South-Central Indiana.** *Solid, interior forests; dotted, forest edge (0 to 200 meters); white, nonforest.*

needed throughout the region. Noss (1983) provided a good overview of the complicated nature of managing highly fragmented, human-dominated landscapes typical of the Central Hardwood Region.

Although past research has provided an understanding of forest and wildlife management for individual sites and for selected species, more research is now needed on the management of forested landscapes at various spatial scales and on the long-term maintenance of mixed-species populations in this and other regions of the United States (Burley 1989; Soule and Kohm 1989).

A REGION UNIQUE FROM OTHER FOREST REGIONS

The social and biological characteristics of the Central Hardwood Region make it unique from other forest regions of the United States. Therefore, issues and management systems being adopted for other regions of the United States are not necessarily appropriate for central hardwoods. The region contains one-fourth of the U.S. human population, with about 90 percent of the land in private ownership. Private lands tend to be in small ownerships, and forest industry is much less important than in other regions. Public lands are also relatively small (maximum of 1.5 million acres with most <100,000 acres) and are usually quite fragmented with private inholdings.

Most of the national forests within the region own less than 50 percent of the land within their purchase boundaries (table 7.1). This compares to 90 percent ownership within the Pacific Northwest Region (Land Areas of the National Forest System 1990). National forests, the largest public land areas, are located within the major remaining forested parts of the region and are therefore important for the long-term sustainability of the system.

Forest management systems in this region tend to be extensive rather than intensive, with very little site preparation after harvest (Mills, Fischer, and Reisinger 1987). Therefore, disruption of plant and animal populations on harvested sites is primarily through the influx of pioneer species and soil disturbance on skid trails. This situation differs from that in other regions, where the site may be burned or windrowed and planted after harvest (Hansen et al. 1991; Walstad, Newton, and Gjerstad 1987).

Group selection and clear-cutting have been the most common regeneration systems used in the region. Group selection is usually combined with single-tree regeneration harvests and has been the most common regenera-

TABLE 7.1.   **Land Ownership within Purchase Boundaries of National Forests of the Central Hardwood Region in 1991**

National Forest	State	Total Land Area (Acres)	National Forest Land Area (Acres)	% National Forest Area
Daniel Boone[a]	KY	1,360,692	525,226	38
Cherokee[a]	TN	1,204,847	626,744	52
Hoosier	IN	643,659	188,330	29
Mark Twain	MO	2,934,454	1,458,299	50
Monongahela	WV	1,650,951	895,624	54
Ozark[a]	AR	1,496,920	1,123,079	75
Shawnee	IL	711,748	256,851	36
Wayne	OH	832,147	202,751	24
Mean for region		1,354,427	659,613	49

[a] 1990 values.

tion system used on private lands and state forests. Clear-cut openings (primarily used on national forests) are relatively small, usually less than 10 hectares in area, and dispersed across the forest. Many of the national forests within the region are now emphasizing group selection with maximum openings of about 2 hectares.

Small, nonindustrial private forest lands are generally poorly managed and are currently being mined because of high stumpage prices (Nyland 1992). Most are harvested with little concern for regeneration or ecosystem sustainability. Only a small percentage of these private owners receive professional advice on forest management. However, there is no widespread conversion of natural forest to single-species plantations within the region as occurs in other regions such as the Pacific Northwest or the Southeast. Plantations usually result from reforestation of agricultural lands.

Biologically, the central hardwood forest ecosystem is quite diverse and very resilient. These forests have survived very severe historical disturbance with limited permanent loss of species. Although there are long lists of state threatened and endangered species, little good quantitative information exists on their historical populations. Reasons for their current rarity vary for each species. Some species may still be recovering from past habitat abuse, but most are rare because of their need for specialized habitat conditions that may occur infrequently across the region. Others may be on the edge of their range within a given state.

Some species are rare because of the current fragmentation of the landscape and interactions with human beings. Many are rare because of a reduction in periodic disturbance, such as fire, that reduces competition with other species (Anderson and Schwegman 1991). Bowles et al. (1990) found that more than half of the endangered and threatened species examined at the Indiana Dunes National Lakeshore responded positively to anthropogenic disturbances.

Although more research is needed on the life history requirements of rare species (Menges 1986), there is little scientific evidence that species (plant or animal) are permanently lost from the site under current forest harvest practices in the Central Hardwood Region. Disturbances currently placed on the system through forest management activities are not nearly as severe as those that have occurred historically. Current natural regeneration systems, including clear-cutting, usually result in the conversion of oak-dominated forests to mixed-species forests (George and Fischer 1992).

Regrowth of forest communities after harvest is rapid across the region, since most species are capable of asexual reproduction or regenerate from stored seed pools. Muller (1982) found similar size class structure and tree species composition in an old-growth and in a 35-year-old second-growth forest that originated from clear-cutting. Both forests had uneven-aged size class distributions. Alpha diversity was greatest in the second growth, and beta diversity was greater in the old-growth forest.

Old-growth conditions develop at 150 to 200 years of age in these forests, and the maximum life span of most tree species is between 350 and 500 years (Parker 1989; McGee 1984). Many of the remaining old-growth remnants of the region are in transition from domination by seral species such as oaks and hickories to more shade-tolerant species such as maples and beech (McCune and Menges 1986). These forests tend to be small fragments scattered as isolated patches across the landscape. Most of these forests were protected by private owners, and many have been lightly disturbed historically through fire, grazing, and partial cutting. Most of the known old-growth remnants are currently protected by private or public organizations.

In summary, the Central Hardwood Region is a highly disturbed landscape resulting from human activities over the last four hundred years.

Relatively high alternative use values, small ownerships, and a large human population have left the forest highly fragmented. A recent shift away from frequent, anthropogenic disturbances is causing a change in forest structure in favor of shade-tolerant species. Old-growth forests that originated from Native American activities in the seventeenth and eighteenth centuries are rare and mostly protected by private or public agencies. The current landscape context of this region precludes the restoration of pre-European conditions, such as large predators or large-scale fire, and will therefore require continued manipulation by humans for long-term sustainability.

THE LANDSCAPE SPATIAL SCALE THAT AFFECTS LONG-TERM SUSTAINABILITY

Much of the debate over the long-term sustainability of this region has revolved around forest management practices on public lands, particularly national forests. This debate has resulted in a shift toward lower intensity silvicultural practices (from clear-cutting to single tree or group selection) and the closure of significant portions of the forests to timber management (to provide core areas for the protection of biological diversity and/or old growth). The emphasis on forest regeneration systems is unfortunate, since it has diverted attention from more important issues at the landscape scale, such as fragmentation, that affect the long-term sustainability of these forests.

Extending rotation lengths, changing the spatial pattern of harvest openings, and reducing road density are probably the most beneficial changes that can be made on public forests managed for various products. Ephemeral openings due to forest harvest are not a threat to the long-term survival of species (such as neotropical birds) if properly arranged in space and time (Thompson and Fritzell 1990). Consideration should be given to concentrating harvest openings such that relatively large (100-hectare) regeneration areas are created to grow into large blocks of mature forest in 150 to 200 years. These regeneration areas could be created through harvest of 10-hectare openings over a 10- to 20-year period. This harvest pattern would slow overall forest fragmentation and create large areas of mature forest for the survival of edge-sensitive species (Franklin and Forman 1987). Road density could also be reduced under this harvest pattern.

Shifting to uneven-aged silvicultural systems may result in reduced diversity and productivity of central hardwood forests because of a more rapid shift to shade-tolerant tree species and consequent decline in oak species (Lamson and Smith 1991). Seral species such as the oaks, walnuts, and hickories are very important in the long-term diversity and productivity of central hardwood forests. These species are a major food source for a wide array of species in the ecosystem.

Uneven-aged silvicultural systems may result in more frequent, less-intense disturbance, greater fragmentation, and more roads than with even-aged systems. Forests harvested with uneven-aged systems require periodic harvest every ten to twenty years to maintain the desired diameter distribution, whereas small clear-cuts may be left undisturbed until the next harvest. Assuming the same volume of harvest, uneven-aged systems create more edge and disturb much larger areas than do even-aged systems. Roads also extend over larger areas and are used more often with uneven-aged systems.

Central hardwood forests managed for multiple products should be managed with both even- and uneven-aged regeneration systems depending on site characteristics and management objectives to ensure their long-term sustainability. Management systems appropriate for specific sites should improve as ecological classification systems, currently being developed, become operational for the region.

Unique sites with adequate buffer areas and old-growth core areas should be protected at least until more is known about species requirements and improved landscape management systems are developed. White-tailed deer populations are expanding in the region (Nixon 1970; Miller, Bratton, and Hadidian 1992) and will require management on protected areas to avoid damage to other species. Putnam (1988) discussed an approach to examining and maintaining vertebrate species diversity in the Mark Twain National Forest in Missouri.

Although all spatial scales are important in the development of sustainable management systems (Wiens 1989; Urban, O'Neill, and Shugart 1987), the major problem affecting the long-term sustainability of the central hardwood forest ecosystem is landscape fragmentation. This is true for both long-term ecosystem integrity and management. The size of contiguous forest areas and consequent distance from agricultural and urban areas are a more serious problem in this region than are forest management activities (Ambuel and Temple 1983). Current trends in development are likely to fragment the system further unless more comprehensive planning is done at the landscape scale. Grumbine (1990) provided a good discussion of the need for landscape-level planning for long-term sustainability of biodiversity.

External edge due to change in land use (forest to agriculture or urban) is believed to be a more serious problem for edge-sensitive species than is the temporary internal edge created through forest harvest. Private inholdings, which lead to the creation of permanent edge, also hinder the development of ecologically sound management systems. For example, instituting the cutting patterns described above and using fire as a management tool will be more difficult under the current land ownership pattern in the region.

Scientific understanding of landscape design to ensure long-term survival of biodiversity is incomplete (Simberloff 1988; Turner 1989), particularly in regions where human activities are the major disturbances. However, several general recommendations can be made to improve forest ecosystem sustainability in the central hardwood landscape over the long-term. First, public (state and federal) forest lands should be consolidated and expanded (inholdings purchased) to ensure that large contiguous areas of the landscape remain in a forested condition; this will enhance the survival of species with large area requirements and the development of ecologically sound management systems. Additional public lands should be purchased and reforested, particularly in the northern half of the region, to provide larger contiguous habitats than are currently present. Incentives to private forest landowners should be continued to improve forest management practices and expand current forest acreage.

Second, since the size of any given contiguous forest area within the region will remain relatively small compared to areas in other regions, a system of connecting corridors should be developed (Harris 1988; Bennett 1990). This is especially true if the large, wide-ranging mammal species are to be reestablished in the central parts of the region. Probably the easiest corridors to develop would be through reforestation of riparian zones along streams and rivers, which would also provide benefits such as improved water quality.

Coupling of state and federal forests within local areas of the region will increase their ecological value (Noss 1983). For example, the largest forest area of Indiana occurs in the south-central region of the state. Public lands in this area are managed as state parks, state forests, state fish and wildlife areas, federal military lands, and national forest lands. The strategic purchase of private lands within the area could connect these public holdings to increase their biological effectiveness. Noss (1987) proposed such a system of connected public lands for southeastern Ohio.

Continued restoration of the central hardwood landscape could be accomplished over the next fifty to one hundred years through a willing seller/willing buyer program with proper long-term planning. Although such a landscape should include an increase in old-growth forest area and protection of unique habitats, most of the forest land in the Central Hardwood Region can be managed for multiple products while providing for long-term ecosystem sustainability.

REFERENCES

Ambuel, B., and S. A. Temple. 1983. Area-dependent changes in the bird communities and vegetation of southern Wisconsin forests. *Ecology* 64:1057-1068.

Anderson, R. C., and J. E. Schwegman. 1991. Twenty years of vegetational change on a southern Illinois barren. *Natural Areas Journal* 11:100-107.

Bennett, A. F. 1990. Habitat corridors and the conservation of small mammals in a fragmented forest environment. *Landscape Ecology* 4:109-112.

Bowles, M. L., M. M. DeMauro, N. Pavlovic, and R. D. Hiebert. 1990. Effects of anthropogenic disturbances on endangered and threatened plants at the Indiana Dunes National Lakeshore. *Natural Areas Journal* 10:187-200.

Burley, J. B. 1989. Multi-model habitat suitability index analysis in the Red River valley. *Landscape and Urban Planning* 17:261-280.

Campbell, J. J. N. 1987. Gradients of tree species composition in the central hardwood region. *Proceedings of the Central Hardwood Forest Conference IV* 6:325-345.

Clark, F. B. 1989. The central hardwood forest. In *Central Hardwood Forest notes*, ed. USDA Forest Service. St. Paul: North Central Forest Experiment Station.

DeVivo, M. S. 1990. Indian use of fire and land clearance in the southern Appalachians. In *Fire and the environment: Ecological and cultural perspectives, proceedings of an international symposium*, 306-310. Gen. Tech. Rep. SE-60. Southeastern Forest Experiment Station.

Franklin, J. F., and R. T. Forman. 1987. Creating landscape patterns by forest cutting: Ecological consequences and principles. *Landscape Ecology* 1:5-8.

George, D., and B. C. Fischer. 1992. The occurrence of oak reproduction after clearcut harvesting on the Hoosier National Forest. *Northern Journal of Applied Forestry* 8:144-146.

Grumbine, E. 1990. Protecting biodiversity through the greater ecosystem concept. *Natural Areas Journal* 10:114-120.

Hansen, A. J., T. A. Spies, F. J. Swanson, and J. L. Ohmann. 1991. Conserving biodiversity in managed forests. *BioScience* 41:382-392.

Harris, L. 1988. Landscape linkages: The dispersal corridor approach to wildlife conservation. *Transactions of the North American Wildlife Natural Resources Conference* 53:595-607.

Ladd, D. 1991. Reexamination of the role of fire in Missouri oak woodlands. In *Proceedings of the Oak Woods Management Workshop,* 67-80. Charleston, IL: Eastern Illinois University.

Lamson, N. I., and H. C. Smith. 1991. Stand development and yields of Appalachian hardwood stands managed with single tree selection for at least 30 years. Research Paper NE-655. Radnor, PA: USDA Forest Service.

Land Areas of the National Forest System. 1990. FS-383. Washington, DC: USDA Forest Service.

Leopold, D. J., G. R. Parker, and W. T. Swank. 1985. Forest development after successive clearcut in the Southern Appalachians. *Forest Ecology and Management* 13:83-120.

Martin, W. H. 1989. *The role and history of fire in the Daniel Boone National Forest.* Winchester, KY: USDA Forest Service, Daniel Boone National Forest.

McCord, S. S. 1970. *Travel accounts of Indiana 1679-1961.* Indianapolis, IN: Indiana Historical Bureau, Indiana Library and Historical Board.

McCune, B., and E. S. Menges. 1986. Quality of historical data in midwestern old-growth forests. *American Midland Naturalist* 116:163-172.

McGee, C. E. 1984. Heavy mortality and succession in a virgin mixed mesophytic forest. Research Paper SO-209. New Orleans: USDA Forest Service.

Menges, E. S. 1986. Predicting the future of rare plant populations: Demographic monitoring and modeling. *Natural Areas Journal* 6:13-25.

Merritt, C. 1980. The Central Region. In *Regional silviculture of the United States,* ed. J. W. Barrett, 107-143. New York: John Wiley & Sons.

Miller, S. G., S. P. Bratton, and J. Hadidian. 1992. Impacts of white-tailed deer on endangered and threatened vascular plants. *Natural Areas Journal* 12:67-74.

Mills, W. L., B. C. Fischer, and T. W. Reisinger. 1987. Upland hardwood silviculture: A review of literature. Station Bulletin No. 527. West Lafayette, IN: Agricultural Experiment Station, Purdue University.

Muller, R. N. 1982. Vegetation pattern in the mixed mesophytic forest of eastern Kentucky. *Ecology* 63:1901-1917.

National Outdoor Recreational Supply Information System. 1987. Athens, GA: USDA Forest Service.

Nixon, C. M. 1970. Deer populations in the Midwest. In Whitetailed deer in the Midwest, 11-18. Research Paper NC-39. St. Paul: USDA Forest Service, North Central Forest Experiment Station.

Noss, R. F. 1983. A regional landscape approach to maintain diversity. *BioScience* 33:700-705.

⸻. 1987. Protecting natural areas in fragmented landscapes. *Natural Areas Journal* 7:2-13.

Nuzzo, V. A. 1986. Extent and status of midwest oak savanna: Presettlement and 1985. *Natural Areas Journal* 6:6-36.

Nyland, R. D. 1992. Exploitation and greed in eastern hardwood forests. *Journal of Forestry* 90:33-37.

Parker, G. R. 1989. Old-growth forests of the Central Hardwood Region. *Natural Areas Journal* 9:5-11.

Parker, G. R., and G. T. Weaver. 1989. *Ecological principles: Climate, physiography, soil, and vegetation.* Central Hardwood Forest Notes. St. Paul: USDA Forest Service, North Central Forest Experiment Station.

Pursell, F. A. 1989. Long-term change in tree populations and the distribution of plants in the edge of woodlands in eastern Indiana. Master's thesis, Purdue University, West Lafayette, IN.

Putnam, C. 1988. The development and application of habitat standards for maintaining vertebrate species diversity on a national forest. *Natural Areas Journal* 8:256-266.

Reeves, M. C. 1976. Wildlife and its management in Indiana from 1716-1900. In *Fish and wildlife in Indiana, 1776-1976: Proceedings of the American Fisheries Society and the Wildlife Society*, 2-4. West Lafayette, IN: The Wildlife Society.

Sander, I., C. Merritt, and E. H. Tyron. 1981. Oak-hickory. In *Choices in silviculture for American forests*, 23-28. Washington, DC: Society of American Foresters.

Simberloff, D. 1988. The conservation of population and community biology to conservation science. *Annual Review of Ecology and Systematics* 19:473-511.

Soulé, M. E., and K. A. Kohm. 1989. *Research priorities for conservation biology.* Washington DC: Island Press.

Steyermark, J. A. 1959. *Vegetational history of the Ozark Forest.* Columbia, MO: University of Missouri Studies.

Thompson, F. R., and E. K. Fritzell. 1990. Bird densities and diversity in clearcut and mature oak-hickory forest. Research Paper NC-293. St. Paul: USDA Forest Service, North Central Forest Experiment Station.

Turner, M. G. 1989. Landscape ecology: The effect of pattern on process. *Annual Review of Ecology and Systematics* 20:171-197.

Urban, D. L., R. V. O'Neill, and H. H. Shugart, Jr. 1987. Landscape ecology: A heirarchical perspective can help scientists understand spatial patterns. *BioScience* 37:119-127.

Walstad, J. D., M. Newton, and D. H. Gjerstad. 1987. Overview of vegetation management alternatives. In *Forest vegetation management for conifer production*, 157-200. New York: John Wiley & Sons.

Ward, J. S., and G. R. Parker. 1989. Spatial dispersion of woody regeneration in an old-growth forest. *Ecology* 70:1279-1285.

Wiens, J. A. 1989. Spatial scaling in ecology. *Functional Ecology* 3:385-397.

8

The Need and Potential for Ecosystem Management in Forests of the Inland West

Robert D. Pfister
School of Forestry, University of Montana

HUMANITY AND THE ENVIRONMENT

The manner in which we relate to each other and to our environment is critical to any definition of sustainability. American society has evolved from a pioneer/settlement society to a highly structured, competitive, technical society in a few hundred years (one hundred years in the western United States). Global population has grown and will grow in an exponential fashion. Depletion of resources is a natural fear during a period of such rapid change. As the threat of nuclear catastrophe declines, more attention is being focused on other threats to our global environment. The inevitable collision of population growth and the equitable distribution of resources urges a serious discussion of alternative futures for the human race on planet Earth. Defining forest sustainability is part of the larger discussion of sustainable development and sustainable societies occurring at the highest echelons of world governments.

Primitive civilizations functioned in a degree of balance with their natural environment (although this balance was not without major fluctuations and hardships for human populations [Chase 1987, 92-115]). In many cases, technology and intensive agriculture have permitted increases in the carrying capacity of the environment and a more "advanced" life style. In many cases, however, technology and population growth have been accompanied by degradation of long-term carrying capacity.

The attitudes, values, and philosophy underlying social actions are central to long-term concerns for sustainability. The Native American philosophy of human-nature harmony is often cited as a useful guiding philosophy for today. However, the Native American philosophy is difficult to relate to today's technologically advanced civilization. Our current population and standards of living are not compatible with that philosophy.

Kimmins (1987) described two prevalent philosophies of today. The first is the traditional or technological philosophy, which views humans as users of natural resources. This philosophy has created problems when use, and especially exploitation, are the major driving forces. A second backlash philosophy has developed as an alternative. It also views humans as separate from their environment but seeks to protect the environment by isolating humans from as much land as possible and seeking environmental protection through legal regulations.

These three philosophies have many champions, advocates, and lobbyists, but none is adequate to address a sustainable future for human society as we know it today. We cannot return to the past, so we must strive to predict and influence the future.

Along with others, Kimmins (1987) proposed a new philosophy, where human society functions as an integral part of natural ecosystems. This has potential for leading toward a sustainable society and sustainable ecosystems—maintaining ecosystems with the long-term capacity to support finite numbers of people at socially defined and regulated standards of living. This new, natural philosophy, however, will probably be slow to catch on, as man's relationship with nature has been debated without resolution for over two thousand years (Chase 1987, 176-177).

The environmental movement, launched in the late 1960s and resurgent in the 1990s, has fostered a misguided reliance on ecological studies to provide direct answers for environmental issues. Unfortunately, this has led in some cases to hostility toward environmental viewpoints, especially when they are pursued with more emotion than fact through political advocacy processes. Kimmins (1987) stated that the ecosystem is the most important concept in ecology and is central to the future of forest management. However, he provided an appropriate context for the discussion of an ecosystem approach relative to sustainable forestry: "Quite incorrectly, ecology has been advanced by some as the basis on which socioeconomic decisions should be made. This is incorrect because the science of ecology itself provides no basis for value judgment and it is only one, albeit of preeminent importance, of several sources of information contributing to socioeconomic decisions" (p. 30).

Recognizing that ecology or ecosystem approaches can provide only part of the solution to questions of sustainability, let us examine what we can gain from an ecological perspective.

THE ECOSYSTEM CONCEPT

The term *ecosystem* was first formally proposed by the English ecologist Tansley (1935) as "not only the organism-complex, but the whole complex of physical factors forming what we call the environment." Other notable ecologists expanded upon the concept to the point where *ecosystem ecology* is now used to describe courses and studies that pertain to the holistic viewpoint—emphasizing that the whole is more than the sum of the interconnected parts.

Kimmins (1987) stated that "it is clear from these definitions that the term ecosystem is more of a concept than a real physical entity—a concept with six major attributes." These six attributes form a foundation for understanding ecosystems and for the contemplation of ecosystem approaches to management.

1. *Structure* consists of biotic and abiotic subcomponents, including both composition and the arrangement of components.
2. *Function* refers to the constant exchange of matter and energy between the physical environment and the living community.
3. *Complexity* results from the high level of biological integration in ecosystems. Events are multiply determined and difficult to predict without knowledge of structure and functional processes.
4. *Interaction and interdependency* of the various living and nonliving components ensures that a change in any one component will result in a change in almost all of the others.
5. *No inherent definition of spatial dimensions* is provided or required by the ecosystem concept. Boundaries and scales are diffuse and multitiered. Geographic boundaries have evolved only through practical application of the ecosystem concept to real landscapes through the development of classification and mapping systems.

6. *Temporal change* is inherent in ecological systems. In addition to the continuous exchanges of matter and energy, the entire structure and function of an ecosystem undergoes change over time.

An ecosystem approach to forest management must address all six attributes of the ecosystem concept.

Structure is essential for communication and documentation of components and relationships among them. *Function* is essential for gaining understanding of processes that are inherently complex and interconnected. *Complexity* is important in that we must accept that decisions will continue to be made with incomplete knowledge for a wide variety of ecosystems. *Interaction and interdependency* are infinitely more complex than our current models are capable of explaining. Since there is *no inherent definition of spatial dimensions* within the concept, it becomes necessary to use our knowledge of structure and composition to develop ecological classification systems that provide a framework for using ecological concepts on real landscapes—moving from the abstract to the visible application of the concept. The attribute of *temporal change* is essential for two reasons. First, it fosters the acceptance of the dynamic nature of ecosystems over time and the futility of conservation attempts that seek to maintain the status quo. Second, it necessitates learning to predict future consequences of alternative management activities.

THE NEED FOR AN ECOSYSTEM APPROACH

The need for an ecosystem approach is essential in forests of the Inland West, the United States, and the world, though certainly it is not universally accepted. Often, it is viewed as a means to stop development, to stymie economic activity, to disrupt existing social systems, and, in some people's minds, simply to stop the harvest of trees. Unless the need for an ecosystem approach is presented objectively and within a rational social context, it will not be accepted by those who have experienced direct threats to their livelihood and security and to the well-being of their families.

Rowe (1992) presents a timely discussion of what the "ecosystem approach" really means, and concludes with the following statement: "The ecosystem approach is a new way of sensing the world, a reevaluation of the Ecosphere and of our place in it, an ethical way of living in and with it. Were the concept more widely shared, a great many more problems than those of foresters and other resource managers co-operating would be solved."

The hope for recognition of the validity of an ecosystem approach is through both research and education aimed at a true understanding of ecology and ecosystems. The facts must be presented clearly without the advocacy of self-aggrandizing egos and the media-mania of society and politics.

THE FOREST ECOSYSTEMS OF THE INLAND WEST

Forest ecosystems in the Inland West have been classified in two fundamentally different ways: (1) existing vegetation, or forest cover type, and (2) potential vegetation in the absence of disturbance. Forest cover types have been the standard in forest inventory for many years and are described by the Society of American Foresters (Eyre 1980). The major forest types in the Rocky Mountains are shown in table 8.1. Of a total of

TABLE 8.1. **Forest Cover Types of the Rocky Mountain States**

Cover Type	Total Area (Million Acres)
Chaparral	7
Pinyon-juniper	42
Ponderosa pine	18
Douglas fir	15
Western larch	2
True firs	7
Hemlock-cedar	1
Lodgepole pine	17
Spruce-fir	9
Other softwoods	6
Aspen	7
Cottonwood/riparian	2

From Green and Van Hooser (1983).

138 million acres, the largest types in order are pinyon-juniper, ponderosa pine, lodgepole pine, Douglas fir, spruce-fir, and true firs. The major hardwood type is aspen (Green and Van Hooser 1983). Other types of limited extent are nevertheless important on a regional basis, such as western white pine, western larch, and cedar-hemlock. This classification is adequate for describing existing dominance, often by seral species, but is inadequate for an ecosystem approach because of wide variation in environments and temporal instability of many seral types.

A complementary method of ecosystem classification has been developed to provide a framework for ecological understanding. Potential natural vegetation classification, including both "potential associations" and "habitat types," has been developed during the past three decades to provide a taxonomic classification for virtually all of the forest ecosystems in the Inland West. This classification approach is often considered to be an alternative to cover types; in reality, however, ecosystems cannot be discussed without combining both approaches (table 8.2). Knowledge of both current and likely future conditions is essential. Since cover types span a broad range of environments and habitat types can include a large number of cover types as different successional stages, a combination provides a level of precision and understanding that cannot be achieved by either alone.

Three major kinds of ecosystems (spruce-fir, lodgepole pine, and ponderosa pine/Douglas fir) will be discussed briefly to illustrate the need and potential for an ecosystem approach in contrast to generalized practices of development or preservation.

The use of these three examples follows a conventional written approach to communicate knowledge relative to different kinds of forest communities and sites. However, an ecosystem approach to management in the real world requires recognition of a mosaic of types on variable landscapes. This is especially evident in the mountains, where a single watershed may have annual precipitation of 10 inches in the lower reaches and 100 inches in the upper reaches, reflected by a wide array of ecological types. Integrating this diversity in forest planning is essential for an ecosystem approach to management. The emerging science of landscape ecology (Zonneveld and Forman 1989) will be a valuable tool for an ecosystem approach to sustainability, as is illustrated by other authors in this section.

TABLE 8.2. **General Distribution of Forest Cover Types across Different Potential Natural Vegetation Types (Indicated by Climax Dominant Tree Species)**

Current Forest Cover Type	Potential Natural Vegetation (Habitat Types)				
	Ponderosa Pine	Douglas Fir	Grand Fir-White Fir	Cedar-Hemlock	Subalpine Fir
Ponderosa pine	X	X	X	X	
Douglas fir		X	X	X	X
Western larch		X	X	X	X
True firs			X	X	
Hemlock-cedar				X	
Lodgepole pine		X	X	X	X
Engelmann spruce-subalpine fir			X	X	X
Aspen		X	X	X	X

SPRUCE-FIR ECOSYSTEMS

Spruce-fir cover types occur as seral and near-climax stands in the subalpine zone, where subalpine fir is the most shade-tolerant conifer and the usual climax dominant (although often sharing that relationship with the spruce). Seral species in addition to Engelmann spruce include lodgepole pine, Douglas fir, and whitebark pine. Environments are cold, with heavy accumulation of snowpack and a short growing season. The predominant natural agents of change are insects, disease, and infrequent, stand-replacing wildfires.

Heavy partial cutting in spruce-fir ecosystems in the northern Rocky Mountains occurred in conjunction with widespread bark beetle epidemics in the old-growth spruce cover types in the 1950s and 1960s. Managers were encouraged to use even-aged management techniques to remove dead and dying spruce (valuable species) and to replace the associated subalpine fir (low-value species) with young stands of spruce. A published research summary recommended clear-cutting in narrow strips with natural regeneration or planting after burning or scarification (Roe, Alexander, and Andrews 1970). This was based on a summary of research and survey studies throughout the northern and central Rocky Mountains. However, regeneration in the southern parts of the type was inadequate using the recommended techniques (Fiedler, McCaughey, and Schmidt 1985). Ecological classification studies (Pfister 1972), age-structure analyses (Hanley, Schmidt, and Blake 1975), and regeneration experiments in Colorado and Utah (Alexander 1984; Pfister 1973) demonstrated that the same cover type had a wide range of functional processes leading to similar-appearing stands. Whereas the natural dynamic process of old-growth senescence and fuel build-up, stand-replacing wildfire, and natural regeneration of even-aged stands was the general process in northern areas, many of the southern stands rarely had fire because of frequent summer precipitation; stand replacement was a gradual, continuing process (gap replacement). Recognition of wide variation within the spruce-fir cover type has led to changes in prescriptions and management embodying a full range of silvicultural practices for different ecosystems and objectives (Alexander 1986, 1987). Alexander (1974) also emphasized that future research should be focused on reevaluation of silvicultural knowledge within a habitat type classification framework.

This example illustrates the need for spatial resolution through ecological classification (see chapter 6) and the importance of recognizing variation in both function and dynamic processes among different ecosystems. An ecosystem approach will require custom design of silvicultural prescriptions and management practices for local ecosystems and stands. We must continue to resist both standard prescriptions and prescription by regulation; either approach (although based on good intentions) will prevent the consideration of many positive alternatives.

LODGEPOLE PINE ECOSYSTEMS

The lodgepole pine ecosystem consists primarily of seral forests extending over large areas within the northern Rocky Mountains and the Inland West. Many of the forests are in the process of being slowly replaced by more shade-tolerant associates such as Douglas fir, true firs, and Engelmann spruce. Replacement at midelevations and on more productive sites may take place in less than a century; at higher elevations, however, the natural replacement by shade-tolerant species may take two or three centuries. The successional process is accelerated at lower elevations by early mortality from bark beetles.

The Yellowstone fires of 1988 brought the ecology and management of lodgepole pine ecosystems to the forefront of national attention. Several ecologists (Arno 1980; Roe and Amman 1970; Despain and Sellers 1977; Romme 1982) had documented the natural role of fire in the development and maintenance of lodgepole pine ecosystems, so the event was no surprise to scientists.

An ecosystem approach to fuels and fire management was being pursued in the 1980s but was not fully in effect at the time of nature's "catastrophic" triumph. The scale and intensity of the Yellowstone fires was largely unacceptable to the public and has contributed to a strong social resistance to the use of fire in natural forest ecosystems. Social acceptance of natural fire will require an appreciation for the importance of these events to ecosystem function. A postfire workshop evaluation report stated that,

> rather than ecological disasters or catastrophes, high-intensity fires in ecosystems such as those of the greater Yellowstone area are virtually inevitable, and are even essential for the successful reproduction of some species. The processes that have regulated ecosystems in the greater Yellowstone area for millennia will continue to operate. The complex mosaic created by the 1988 fires will initiate successional processes that will guarantee future landscape variety and diversity. (Christensen et al. 1989)

The lodgepole pine ecosystem is a classic example of a seral ecosystem that requires major natural disturbance processes at fairly regular intervals. The natural seral ecosystem cannot be maintained by traditional approaches to preservation; the system must be periodically disturbed. The lack of young, healthy lodgepole pine stands (after stand replacement fires) is specifically of great concern relative to the survival of the Canadian lynx in eastern Washington (Chasan 1991). A change in social attitudes is needed before a more natural ecosystem approach to management will become socially acceptable.

PONDEROSA PINE/DOUGLAS FIR ECOSYSTEMS

Ponderosa pine and Douglas fir cover types are extensive throughout the Inland West. Ponderosa pine occurs as a relatively pure, climax type at the lowest forest elevations and as a seral cover type at midelevations, where the common shade-tolerant, climax associates are Douglas fir and grand fir. Rainfall ranges from 15 to 30 inches per year in these warm, dry environments. Natural presettlement fire frequencies in the pure stands were five to thirty years, generally as underburns with minimal effects on larger trees—thereby maintaining a parklike stand structure (Arno 1980).

An infamous example of the need for an ecosystem approach to management in this type is the Bitterroot National Forest in western Montana. Fire exclusion and selective harvesting of ponderosa pine and western larch through the 1950s created many stands of unhealthy, fire-susceptible, shade-tolerant Douglas fir. These were viewed as degraded ecosystems by forest managers concerned with maintaining healthy, productive forests for sustained yield. A dramatic move to even-aged management was made in the 1960s to perpetuate ponderosa pine as a featured species of management. Clear-cutting, often with terracing and planting, became a favored timber management technique—that is, until strong opposition arose from the public. Important concerns were voiced regarding management effects on wildlife, watershed protection, and aesthetics. An agency review team was assembled to review the management situation (Worf et al. 1970). An independent review team from the University of Montana was also commissioned to review the management situation (Bolle et al. 1970). Changes were made in both practice and approach as a result of these reviews. However, management controversy continues among a polarized citizenry with strongly contrasting viewpoints of how and for whom the public lands should be managed. Citizens and professional land managers seem to be locked into a dysfunctional relationship, where enormous human resources are being used in a continuing process of conflict and conflict resolution on each and every proposed management action. Ironically, the conflict resolution often ignores basic principles of long-term forest health, thereby creating future undesirable conditions.

Research in silviculture, insects, disease, watershed, wildlife, plant succession, and fire ecology has provided a good knowledge base. However, functionalism and polarization have reached the point where managers are continually seeking new ways to manage the land effectively with local public acceptance. New perspectives in multiresource silviculture and landscape ecology are being pursued to help resolve long-standing conflicts.

Major problems in this ecosystem were also reported from eastern Oregon and eastern Washington by Wickman (1992). Fire exclusion and selective removal of intolerant seral species (ponderosa pine and western larch) have produced multistoried stands of shade-tolerant conifers (Douglas fir and true fir). These stands are being decimated by disease (dwarf mistletoe and root rots) and insects (spruce budworm and tussock moth). Extreme mortality has resulted from creating stand conditions very different from natural, fire-initiated successional processes in this ecosystem. Natural fires either reduced fuel accumulation periodically by under-burning (at average intervals of 10 to 25 years) or replaced existing stands with a stand replacement fire (every 50 to 150 years) that produced seral stands of healthy trees (Mutch et al. 1993). Fire exclusion has set the stage for uncontrollable, severe fires. Restoration will require a careful combination of silviculture and prescribed fire to restore a semblance of sustainable natural forest conditions.

Major obstacles to the restoration of natural processes in these ecosystems include concerns about air quality in relation to prescribed burning; potential risks associated with returning fire to the ecosystem; the need for data to ensure professional credibility; functional resistance regarding aesthetics, water quality, and wildlife by professionals fighting to maintain the status quo; and the need for administrative leadership in risk taking to restore ecological processes (Mutch et al. 1993).

Here again we are faced with a situation where the lack of an ecosystem approach has created undesirable forest conditions. However, an ecosystem approach to management will be difficult to achieve because of strong social and economic constraints.

OTHER ECOSYSTEMS

Changes are taking place in the cedar-hemlock-white pine ecosystems, western larch ecosystems, and others where the presence of resources with high economic value has led to continuing extraction and management activities. Other ecosystems of lower commercial value (e.g., aspen) are progressing through natural dynamic processes (except for exclusion of natural fires), in some cases to their own demise. Change is inherent in both management activity (development) and nonactivity (preservation).

An ecosystem approach to management will need to predict alternative future conditions as a framework to help resolve conflicts in social desires and values.

THE POTENTIAL FOR AN ECOSYSTEM APPROACH

Ecological research of the past few decades has been building a solid foundation for ecosystem management in the Inland West. Many managers are making continual changes in management practices based on results of ecological studies. These can be related back to the attributes of ecosystems presented earlier.

STRUCTURE

Knowledge of structure has been advanced greatly through the development of classification systems. The development of data bases, definition of types, compositional description of types, and delineation of relationships to other ecosystem components have greatly advanced our knowledge of the forest ecosystems of the Inland West (Ferguson, Morgan, and Johnson 1989).

FUNCTION

Studies of nutrient cycling relative to logging and burning practices have been conducted during the past two decades to evaluate environmental effects and long-term site degradation in different ecosystem types (Stark 1982, 1988). New management practices have been recommended for the few situations where cause for concern has been clearly demonstrated. Studies of intensive timber utilization in the 1970s demonstrated the importance of coarse woody debris retention to the maintenance of long-term site quality (Harvey, Larsen, and Jurgensen 1979). National forests in the Northern Region have been requiring a minimum of 12 tons of coarse residues per acre during the past decade, and cooperative studies are continuing to move from this blanket recommendation to a set of "ecosystem type-specific" recommendations. Snag retention for cavity-nesting species has become management policy during the past two decades. Green-tree retention is being evaluated as an alternative to clear-cutting. Silviculturists are reevaluating the full range of silvicultural alternatives relative to the ideas of New Forestry (Franklin 1989) and New Perspectives (Salwasser 1990).

The development of process models has provided new dimensions for understanding the function of ecosystems (Running and Coughlan 1988; Kimmins and Scoullar 1983). Refinement of these models and linkage to spatial models (or classifications) and temporal succession models (or classifications) will lead to new dimensions of understanding and management applications.

INTEGRATION AND INTERDEPENDENCE

Ecological classification efforts necessarily integrate components of ecosystem structure in the development of taxonomies and mapping classes. Additionally, aspects of different classification systems have been further integrated for special applications (Driscoll et al. 1984). Modeling efforts in the area of processes (function) and succession have integrated new information and highlighted ecosystem interdependence. Current efforts in landscape ecology are incorporating a new set of variables related to biodiversity with a wide range of scales and resolutions.

SPATIAL RESOLUTION

Taxonomies for ecological site classification were developed for most forest lands of the Inland West during the 1970s and 1980s. In addition to providing a valuable communication and stratification tool, the classifications have been used to develop type-specific summaries of management knowledge. This has allowed the replacement of generalized blanket and cover type recommendations with site-specific professional evaluation. The classification system has been especially useful to document variation in functional processes in different ecosystems and provides a foundation for extrapolation to appropriate sites (Ferguson, Morgan, and Johnson 1989). Training has increased utilization of the habitat type system; mapping has thus far been completed only for relatively small portions of the Inland West, but many other inventories are cross-referenced to habitat types.

Spatial delineation of ecosystems (mapping) with integration of vegetation, soil, climate, and landform is not fully resolved for various scales in spite of numerous task forces, workshops, symposiums, and publications seeking common ground among contrasting functional and disciplinary perspectives. Management analysis at the landscape level is an emerging field with new training programs, research, development, and pilot demonstration projects. These actions have renewed the debates in the field of ecological land classification dealing with the integration of taxonomic and mapping approaches and the development of operational classification hierarchies. Studies in watershed management, wildlife management, geographic information systems, and landscape ecology are clearly illustrating the problems of scale in putting spatial boundaries on ecosystems as part of a land management process.

TEMPORAL CHANGE

Studies of ecosystem dynamics are at the tantalizing stage where a concerted, well-planned effort would yield confident predictions of future ecosystem conditions. Data bases and knowledge of species responses to disturbances are accumulating as a foundation for new modeling efforts (Arno, Simmerman, and Keane 1985; Fischer 1989; Keane 1989; Moeur and Ferguson 1989). Classification of successional stages and successional pathways in relation to habitat types and various treatments (processes) has been completed on a pilot basis (Arno, Simmerman, and Keane 1985; Steele 1984, 1989). Recent studies that link successional processes to ecological site classifications permit extrapolation to novel scales (Keane, Arno, and Brown 1990). Fire ecology studies, as suggested by Pinchot (1899), have increased our understanding of the prehistoric role of natural and human-caused fire and have forced reconsideration of the effects of various fire management policies on ecosystem dynamics (Arno 1980; Fischer and Bradley 1987; Kessell and Fischer 1981).

We are gaining the ability to predict the consequences of alternative management practices for different kinds of stands and ecological sites. As our ability to predict the future improves, we will have the basis for presenting alternatives to the public in a way that improves understanding of the consequences of alternative courses of action. This may lead eventually to a consensus on desired future conditions of stands and landscapes and the actions necessary to obtain them.

THE INTEGRATION OF ECOSYSTEM ATTRIBUTES

An additional benefit of emphasizing an ecosystem approach to management will be the integration of ecosystem attributes into a holistic understanding of the lands and stands that we manage. Fragmentation is a biodiversity concern, but the term also applies well to the problems of functionalism as a major obstacle to ecosystem research as well as ecosystem management. Ecological studies are often focused on single attributes of the ecosystem concept. Great advances can be made when these studies are integrated.

An excellent example of integrating ecosystem attributes into land management is a recent program of the Northern Region of the USDA Forest Service (1991) entitled "Sustaining Ecological Systems." This program provides regional leadership to establish a solid ecosystem foundation for the planning and management of public lands. A commitment has been made at the regional level, and plans for implementation arc under way.

ECOLOGY IS NECESSARY BUT NOT SUFFICIENT

Finally, the best of ecological approaches cannot sustain ecosystems unless they are integrated into a human context. In addition to being ecologically responsible, ecosystem management must be economically viable and socially acceptable (see chapter 2). Economic viability requires a balance of shared costs and benefits among all members of society, and social acceptance is the controlling factor in any use and management of ecosystems (Wenz 1988). Social acceptance is fundamental to all societies, but it is especially important in a humanitarian society where freedom of expression, equal opportunity, and self-governance are maintained.

ECOSYSTEM MANAGEMENT AND SUSTAINABLE FORESTRY

The search for sustainable forestry is driven by the hope that ecology can provide the solutions to a social dilemma. In its broadest sense, ecology may provide many of the answers—but only if it is holistic enough to incorporate the human element as part and parcel of the ecosystem. Unfortunately, the science of ecology is very immature in terms of integrating the human component as part of the structural attribute of ecosystems.

For humans to live within ecosystems, they must be an active participant in their dynamics; human activity must be integrated with natural process. Social acceptance must be encouraged through an educational process that demonstrates the long-term benefits of sustainable ecosystems and that counterbalances the social fears of unbridled natural process, such as wildfire. People must come to appreciate that preservation of the status quo is not an effective way of living within dynamic ecosystems for the long-term benefits of an orderly integration of social values and natural processes. Dynamic society and dynamic ecosystems dictate that the concept of sustainability must include a form of dynamic equilibrium. Furthermore, forestry and conservation professionals need to reaffirm their commitment to the long-term perspectives upon which their professions were originally based.

Sustainable forestry is an extremely important and timely concept as we move toward the twenty-first century. The topic has been approached indirectly for the past century, but for the most part we have not been able to establish or maintain a clear direction for the future. Short-term constraints override our ability to achieve long-term objectives. We are so concerned with the here and now that we do not take the time to explore alternative futures, much less make the kind of commitment needed to assure the realization of a desired future condition. Part of the problem is merely technical; we lack the ability to forecast future conditions adequately. Another part is our unwillingness as a society to make the short-term sacrifices required to attain a desired future condition. This is evident in our consumption habits and the fiscal policies of our elected officials. As the bumper sticker says, "we are spending our children's inheritance"— in fact, we are even borrowing heavily against it. A responsible society will plan for the future, predicting events as well as possible but adjusting its actions as knowledge evolves. A responsible society will provide, to the best of its ability, a livable, enjoyable environment for future generations. Environmental justice is needed—today and for the future.

As we continue to define sustainable forestry, I hope that we can separate out the general concepts that are applicable throughout the United States, identify a method to move toward sustainability on a local basis, and avoid the temptations to overgeneralize. The best approach to long-term sustainability may be to seek it at local levels and obtain it at the national level through the process of aggregation and integration. A bottom-up approach may be much more effective than a top-down approach. Direction from the national level will be more effective in the long run if it enables and empowers local entities to resolve problems locally while thinking globally.

CONCLUSION

Existing knowledge and continuing research in ecological applications have great potential to provide part of the knowledge base essential for sustainable forestry. It is also necessary to incorporate the human element as part and parcel of the ecosystem—perhaps as part of the science of ecology. However, the solution to the sustainable forestry puzzle will probably not be found in the science of ecology alone, but in a new integration of ecological science, social science, and philosophy.

Two sociologists provided eight "answers" for what forests should sustain, including dominant products (economic), community stability (social), human benefits (social), the global village, and four ecological perspectives (Gale and Cordray 1991). These eight answers help illustrate prevalent competing values and demonstrate the need for social conflict resolution, despite its difficulty.

An additional ecological "answer" should be added to their list—namely, dynamic landscape ecosystem management. This ecological approach is founded upon natural dynamic changes in time and space and the human interaction needed to maintain the dynamics through planned manipulation of forest communities. The emphasis is on preserving the process of change—not on the status quo. If this becomes an acceptable goal of management, it may be easier to resolve many of the nonecological social and economic values that society wishes to sustain. Here again the answer lies not in championing a single answer, but in seeking an acceptable, judicious combination of a multiple set of goals in the spirit of social harmony.

The rapidly developing field of landscape ecology offers considerable promise for documenting the dynamics of landscapes, evaluating land suitability, and integrating ecological, social, technological, and economic perspectives (Zonneveld and Forman 1989). The development of landscape ecology as a science has been evolving in Europe for the past four decades, but in the United States it lay relatively dormant until a new awakening in the 1980s. The science is expanding in many directions and will need to develop further operational applications to land and resource management conflicts.

Although American forestry is often criticized for its roots in European forestry, we could benefit by studying Europe's responses to changing social values over the past twenty years. Europe has developed an integrated, sustainable, multiple-use forest management concept where the land base is too small to segregate different land uses in different areas. The management for each unit must be planned according to natural, political, economic, and social parameters (Koch 1990; Plochman 1989).

Additional direction is provided by Brooks and Grant (1992):

> The new approach is developing from a set of hypotheses about how natural systems operate and appropriate human use of forested ecosystems. . . . Forest management must be based upon an ecosystem perspective . . . that recognizes the need to design practices that are sensitive to the balance among various components of the forest. . . . We reject the notion that the changes and troubles faced by forest managers have emerged only recently and are the product of an unappreciative public stimulated by 'radical environmentalists.' . . . Rather, forest managers must recognize that the findings of forestry science, . . . closely associated with results in other areas of science—are forcing us to rethink our approaches to management.

In the United States, we are discussing "sustainable forestry" while the world is discussing "sustainable development" (Barron and Rotherham 1991; Baskerville 1991; Koch 1990; Maini 1991; Rannard 1991; Squire 1990; World Commission on Environment and Development 1987). Other countries are looking toward a combination of sustainable development and stewardship. Are we in American forestry doing the same, or are we unwilling to become involved in the larger sustainable development debate? Long-term conflict resolution in forestry will not take place until polarized advocates are willing to engage in open discussions on global perspectives and explore the contradictions in sustainable development (Redclift 1987).

In her essay "The Search for Sustainability," Rees (1990) provided a sobering overview of the seriousness of the sustainability challenge. I close with some of her poignant observations.

> The ecology and economics of human development are becoming evermore interwoven—locally, regionally and globally—into a seamless network of causes and effects. . . . These interrelationships . . . require that people in all countries and all walks of life work urgently to restructure national and international policies and institutions in order

to foster the sustainability of social and economic development. . . . There is now a widespread consensus that human-induced changes in the fundamental characteristics of natural life support systems are proceeding at a non-sustainable pace. . . . Responses are still largely reactive. . . . They are doomed to failure. . . . Short-sighted national self-interest still predominates in the international political economy. . . . Until the message of the Brundtland Commission, that we all share a common future, is translated from political rhetoric to economic reality, the search for sustainable development on a global scale will fail. . . . A sustainable development path will be futile unless the distribution of wealth and welfare is finally addressed.

REFERENCES

Alexander, R. 1974. Silviculture of central and southern Rocky Mountain forests: A summary of the status of our knowledge by timber types. Research Paper RM-120. Ft. Collins, CO: USDA Forest Service, Rocky Mountain Forest and Range Experiment Station.

_____. 1984. Natural regeneration of Engelmann spruce after clearcutting in the Central Rocky Mountains in relation to environmental factors. Research Paper RM-254. Ft. Collins, CO: USDA Forest Service, Rocky Mountain Forest and Range Experiment Station.

_____. 1986. Silvicultural systems and cutting methods for old-growth spruce-fir forests in the central and southern Rocky Mountains. Gen. Tech. Rep. RM-126. Ft. Collins, CO: USDA Forest Service, Rocky Mountain Forest and Range Experiment Station.

_____. 1987. Ecology, silviculture, and management of the Engelmann spruce-subalpine fir type in the Central and Southern Rocky Mountains. Research Paper RM-254. Ft. Collins, CO: USDA Forest Service, Rocky Mountain Forest and Range Experiment Station.

Arno, S. F. 1980. Forest fire history in the northern Rockies. *Journal of Forestry* 78:460-465.

Arno, S. F., D. G. Simmerman, and R. E. Keane. 1985. Forest succession on four habitat types in western Montana. Gen. Tech. Rep. INT-177. Ogden, UT: USDA Forest Service, Intermountain Research Station.

Barron, D. E., and T. Rotherham. 1991. Towards sustainable development in industrial forestry. *Forestry Chronicle* 67(2):113-116.

Baskerville, G. L. 1991. Concluding comments. *Forestry Chronicle* 67(2):117-118.

Bolle, A. W., R. W. Behan, W. L. Pengelly, R. F. Wambach, G. Browder, T. Payne, and R. E. Shannon. 1970. A university view of the Forest Service: Report on the Bitterroot National Forest. Senate Document 91–115. Washington, DC: US Government Printing Office.

Brooks, D. J., and G. E. Grant. 1992. New approaches to forest management: Background, science issues, and research agenda. *Journal of Forestry* 90(1):25-28.

Chasan, D. J. 1991. Dying forests. *Seattle Weekly*, Aug. 28:36-43.

Chase, A. 1987. *Playing God in Yellowstone*. San Diego: Harcourt Brace Jovanovich.

Christensen, N. L., J. K. Agee, P. F. Brussard, J. Hughes, D. H. Knight, G. W. Minshall, J. M. Peek, S. J. Pyne, F. J. Swanson, S. Wells, J. W. Thomas, S. E. Williams, and H. A. Wright. 1989. Ecological consequences of the 1988 fires in the greater Yellowstone area. Final report, The Greater Yellowstone Postfire Ecological Assessment Workshop. Yellowstone National Park, Gardiner, MT.

Despain, D. G., and R. E. Sellers. 1977. Natural fire in Yellowstone National Park. *Western Wildlands* 4:20-24.

Driscoll, R. S., D. L. Merkel, D. L. Radloff, D. E. Snyder, and J. S. Hagihara. 1984. An ecological classification framework for the United States. USDA Misc. Pub. No. 1439.

Eyre, F. H., ed. 1980. *Forest cover types of the United States and Canada*. Washington, DC: Society of American Foresters.

Ferguson, D. E., P. Morgan, and F. D. Johnson, compilers. 1989. *Proceedings— land classifications based on vegetation—applications for resource management, 1987 November 17-19, Moscow, ID*. Gen. Tech. Rep. INT-257. Ogden, UT: USDA Forest Service, Intermountain Research Station.

Fiedler, C. E., W. W. McCaughey, and W. C. Schmidt. 1985. Natural regeneration in intermountain spruce-fir forests—a gradual process. Research Paper INT-343. Ogden, UT: USDA Forest Service, Intermountain Research Station.

Fischer, W. C. 1989. The fire effects information system: A comprehensive vegetation knowledge base. In *Proceedings—land classifications based on vegetation—applications for resource management, 1987 November 17-19, Moscow, ID*, 107-113. Gen. Tech. Rep. INT-257. Ogden, UT: USDA Forest Service, Intermountain Research Station.

Fischer, W. C., and A. F. Bradley. 1987. Fire ecology of western Montana forest habitat types. Gen. Tech. Rep. INT-223. Ogden, UT: USDA Forest Service, Intermountain Research Station.

Franklin, J. F. 1989. Toward a new forestry. *American Forests*, Nov./Dec.:1-8.

Gale, R. P., and S. M. Cordray. 1991. What should forests sustain? Eight answers. *Journal of Forestry* 89(5):31-36.

Green, A. W., and D. D. Van Hooser. 1983. Forest resources of the Rocky

Mountain States. Resource Bulletin INT-33. Ogden, UT: USDA Forest Service, Intermountain Forest and Range Experiment Station.

Hanley, D. P., W. C. Schmidt, and G. M. Blake. 1975. Stand structure and successional status of two spruce-fir forests in southern Utah. Research Paper INT-176. Ogden, UT: USDA Forest Service, Intermountain Forest and Range Experiment Station.

Harvey, A. E., M. L. Larsen, and M. F. Jurgensen. 1979. Biological implications of increasing harvest intensity on the maintenance and productivity of forest soils. In *Symposium proceedings: Environmental consequences of timber harvesting in Rocky Mountain coniferous forests, September 11-13, 1979, Missoula, MT*, 211-220. Gen. Tech. Rep. INT-90. Ogden, UT: USDA Forest Service, Intermountain Forest and Range Experiment Station.

Keane, R. F. 1989. Classification and prediction of successional plant communities using a pathway model. In *Proceedings—land classifications based on vegetation—applications for resource management, 1987 November 17-19, Moscow, ID*, 56-62. Gen. Tech. Rep. INT-257. Ogden, UT: USDA Forest Service, Intermountain Research Station.

Keane, R. F., S. F. Arno, and J. K. Brown. 1990. Simulating cumulative fire effects in ponderosa pine/Douglas-fir forests. *Ecology* 71(1):189-203.

Kessell, S. F., and W. F. Fischer. 1981. Predicting postfire plant succession for fire management planning. Gen. Tech. Rep. INT-94. Ogden, UT: USDA Forest Service, Intermountain Forest and Range Experiment Station.

Kimmins, J. P. 1987. *Forest ecology*. New York: Macmillan.

Kimmins, J. P., and K. A. Scoullar. 1983. *FORCYTE-10: A user's manual*. Vancouver: Faculty of Forestry, University of British Columbia.

Koch, N. E. 1990. Sustainable forestry: Some comparisons of Europe and the United States. In *Starker Lectures: 1990. Sustainable forestry: Perspectives for the Pacific Northwest*, 41-53. Corvallis, OR: College of Forestry, Oregon State University.

Maini, J. S. 1991. Practicing sustainable forest sector development in Canada: A federal perspective. *Forestry Chronicle* 67(2):107-108.

Moeur, M., and D. E. Ferguson. 1989. Forecasting secondary succession with the prognosis model and its extensions. In *Proceedings—land classifications based on vegetation—applications for resource management, 1987 November 17-19, Moscow, ID*, 67-73. Gen. Tech. Rep. INT-257. Ogden, UT: USDA Forest Service, Intermountain Research Station.

Mutch, R. W., S. F. Arno, J. K. Brown, C. E. Carlson, R. D. Ottmar, and J. L. Peterson. 1993. Forest health in the Blue Mountains: A management strategy for fire-adapted ecosystems. Gen. Tech. Rep. PNW-GTR-310. Portland, OR: USDA Forest Service, Pacific Northwest Research Station.

Pfister, R. D. 1972. Vegetation and soils in the subalpine forests of Utah. Ph.D. diss., Washington State University, Pullman, WA.

————. 1973. Habitat types and regeneration. In *Western Forestry and Conservation Association Proceedings, Portland, OR, 1972*, 120-125. Portland, OR: WFCA.

Pinchot, G. 1899. The relation of forests and forest fires. *National Geographic* 10:363-403.

Plochman, R. 1989. *The forests of central Europe: A changing view. The 1989 Starker Lectures*, 1-9. Corvallis, OR: College of Forestry, Oregon State University.

Rannard, C. D. 1991. Sustainable development (in forestry)—what does it really mean? *Forestry Chronicle* 67(2):109-112.

Redclift, M. 1987. *Sustainable development: Exploring the contradictions.* London: Routledge.

Rees, J. 1990. Natural resources: Allocation, economics and policy. 2d ed. London: Routledge.

Roe, A. L., R. R. Alexander, and M. D. Andrews. 1970. Engelmann spruce regeneration practices in the Rocky Mountains. Prod. Research Rep. 115. Washington, DC: USDA Forest Service.

Roe, A. L., and G. D. Amman. 1970. The mountain pine beetle in lodgepole pine forests. Research Paper INT-71. Washington, DC: USDA Forest Service.

Romme, W. H. 1982. Fire and landscape diversity in subalpine forests of Yellowstone National Park. *Ecological Monographs* 52:199-221.

Rowe, J. S. 1992. The ecosystem approach to forestland management. *Forestry Chronicle* 68(1): 222-224.

Running, S. W., and J. C. Coughlan. 1988. A general model of forest ecosystem processes for regional applications: I. Hydrologic balance, canopy gas exchange and primary production processes. *Ecological Modelling* 42:125-154.

Salwasser, H. 1990. Gaining perspective: Forestry for the future. *Journal of Forestry* 89(11):32-38.

Squire, R. O. 1990. The forestry profession under siege: Meeting the challenge of balancing sustained wood production and ecosystem conservation in the native forest of South-eastern Australia. In *Starker Lectures: 1990. Sustainable forestry: Perspectives for the Pacific Northwest*, 1-15. Corvallis, OR: College of Forestry, Oregon State University.

Stark, N. 1982. Soil fertility after logging in the northern Rocky Mountains. *Canadian Journal of Forest Research* 12(3):679-686.

————. 1988. Nutrient cycling concepts as related to stand culture. In *Proceedings—future forests of the Mountain West: A stand culture symposium. September 29-October 3, 1988, Missoula, MT*, 210-218. Gen. Tech Rep. INT-243. Ogden, UT: USDA Forest Service, Intermountain Research Station.

Steele, R. 1984. An approach to classifying seral vegetation within habitat types. *Northwest Science* 58(1):29-39.

————. 1989. Pyramid models for succession classification. In *Proceedings—*

land classifications based on vegetation—applications for resource management, 1987 November 17-19, Moscow, ID, 63-66. Gen. Tech. Rep. INT-257.
Ogden, UT: USDA Forest Service, Intermountain Research Station.

Tansley, A. G. 1935. The use and abuse of vegetational concepts and terms.
Ecology 16:284-307.

USDA Forest Service. 1991. Our approach to sustaining ecological systems.
Missoula, MT: USDA Forest Service, Northern Region.

World Commission on Environment and Development. 1987. Our common future: The World Commission on Environment and Development. Oxford: Oxford University Press.

Wenz, P. S. 1988. Environmental justice. New York: State University of New York
Press.

Wickman, B. E. 1992. Forest health in the Blue Mountains: The influence of
insects and diseases. Gen. Tech. Rep. PNW-GTR-295. Portland, OR: USDA
Forest Service, Pacific Northwest Research Station.

Worf, W. A., R. H. Cron, S. C. Troltev, O. L. Copeland, S. B. Hutchinson, and
C. A. Wellner. 1970. Management practices on the Bitterroot National Forest:
A task force appraisal, May 1969-April 1970. Missoula, MT: USDA Forest
Service, Northern Region.

Zonneveld, I. S., and R. T. T. Forman, eds. 1989. Changing landscapes: An
ecological perspective. New York: Springer-Verlag.

9

Ecosystem Management:
An Idiosyncratic Overview

John C. Gordon
School of Forestry and Environmental Studies, Yale University

To succeed as a tool for managing complex systems, ecosystem management (EM) must be describable in a way that can be broadly and readily understood. The description attempted here is a personal one that draws on many sources to adduce five simple descriptors of EM: (1) manage where you are, (2) manage with people in mind, (3) manage across boundaries, (4) manage based on mechanisms rather than algorithms, and (5) manage without externalities.

DEFINITIONS

Ecosystem as used here derives from the physical-chemical definition of system (a bounded space pervious to energy but not mass with transit and transactions of both accounted for). Mass obviously goes in and out of ecosystems in real life but must be accounted for. Ecosystems are usually areas with intrinsic properties. Watersheds or lakes are the most frequently cited examples, but any bounded space can usefully be considered an ecosystem, including Earth to some defined envelope. *Management* implies a human purpose and the means and will to attempt to accomplish it. *Ecosystem management* is, at a minimum, people trying to accomplish something in a bounded space.

People usually try to manage only that which they can influence, so that EM takes place on or from the surface of Earth. EM boundaries are traced on that surface and usually extend upward and downward an unspecified but relatively short distance. The boundary may be flexible in time and with objective, that is, acceptable and useful as long as changes are traced and recorded. Because of its dependence on tracking energy flows, EM is green-plant-centric in forest systems but can be applied to any system whether or not it contains green plants.

EM is now only imperfectly realized anywhere, just like other theoretical forms of management. That should be no barrier to the use of its principles and the practices they imply. EM technology will probably emerge as more important to people than either the technology of the communications revolution or biotechnology because of its potential usefulness in guaranteeing a livable environment.

EM technology will probably be widely applied. Activities like the siting and operation of manufacturing facilities and the administration of large organizations should benefit as much from EM concepts as landscape management, since they involve the manipulation of whole, complex systems.

HISTORY AND CONTEXT

The history of North American land use has set the stage for the emergence of EM. Native Americans vigorously managed portions of the landscape but usually with subsistence rather than the accumulation of a surplus as an objective. Early traders extracted commodities (fish, fur, wood, minerals) but did little that could be described as management. With the great westward migration of European-derived pioneers, land itself became a commodity and an avenue to economic security. Bounding and control of defined portions of the landscape were quickly followed by recognition that the productive capacity of the land was both a criterion of value and manageable. As knowledge and technical capability increased, the manipulation of productive capacity through drainage, fertilization, and improved varieties of plants and animals shaped management methods. The growing recognition that society is ultimately reliant on the landscape for long-term well-being has now called forth the concepts of *sustainability* or *sustainable development* and EM as a tool to achieve sustainability.

The major change in forestry thinking wrought by EM has been the abandonment of the concept of a stable flow of wood from the land as a universally dominant management objective. As an environmental paradigm replaces utilitarian, conservation, and preservation paradigms in land managers' and the public's view of the landscape, the management of whole systems for a variety of purposes rather than for commodity flows or single resources (including "wilderness") will become increasingly overt and explicit. EM will differ from multiple-use management in focusing on inputs, interactions, and processes, as well as on uses or outputs.

SIMPLE ECOSYSTEM MANAGEMENT PRINCIPLES

Complex mathematical and diagrammatic treatments of the principles of EM could and should be assembled and, to some degree, have been. The minimum set of EM instructions, however, includes five simply stated concepts. Their simplicity of statement is misleading because they are easy to say but very hard to do.

MANAGE WHERE YOU ARE

Because EM requires both the recognition and the transcendence of real boundaries, it applies to a specific site. Although the notion of *site specificity* is old in forestry, the realization of EM will require even greater emphasis on the specific properties of the place to be managed and the objectives for which it is managed. Many of the examples of "poor" past practice are results of moving "best practices" out of their effective domain defined by places and objectives.

MANAGE WITH PEOPLE IN MIND

All Earth ecosystems are subject to human influence, known and unknown, overt and accidental. Management itself implies strong human purpose. Indeed, some of the most difficult tasks proposed for EM revolve around the notion of excluding human use and influence from specific systems. Thus, EM begins with a careful evaluation of human desires, influences, and responsibilities. Every system definable in biological and physical terms connects to and interacts with a network of human values, uses, institutions, and other social structures.

MANAGE ACROSS BOUNDARIES

EM tracks transactions across defined system boundaries and moves the boundaries themselves when necessary. At the most elementary level, this requires that "neighbor" influences be recognized and managed. This often must be done without extending fee ownership and through a complex process of joint goal setting, compromise, regulation, and incentives.

MANAGE BASED ON MECHANISMS RATHER THAN "RULES OF THUMB"

Knowledge of the specific processes and interactions responsible for system activities and outputs is the key to continuously improving EM. Landscape-based systems are so complex that they can rarely be managed by enumeration and tracking of all components. Detailed knowledge of why things occur as they do is thus the only path to the prediction of system behavior and output. This requires reliance on sampling and on knowledge gained elsewhere but tested and adapted in the system at hand.

MANAGE WITHOUT EXTERNALITIES

In the sense that it is used here, *externalities* means system contents not currently seen to be related to management objectives. At first glance this seems to contradict the assertion that complete enumeration of system contents is seldom if ever achieved. However, it requires rather that all known contents of the system be included and considered when decisions and manipulations are made.

THE IMPLICATIONS OF ECOSYSTEM MANAGEMENT

Use of just these simple principles consistently would improve the management of most pieces of the landscape. But the supremacy of knowledge that they imply makes EM more difficult and expensive than, say, the management of commodity flows. Specifically, the on-site sampling, both to inventory and monitor a system adequately and to know local mechanisms, is labor intensive and time-consuming. Also, to understand interactions and to integrate the solution to a variety of production and preservation problems require a broad range of technical knowledge and the social skills to assemble and use the people who collectively embody it.

Thus, the central question regarding EM is When can we afford to apply it? This is closely linked to a second major question: By what criteria will we judge its effectiveness? At least a part of the answer to the first is implicit in the answer to the second. The development of objective ways to estimate and track ecosystem health would allow the use of ecosystem health as an effectiveness criterion and as a way to place a value on the output of EM. Components of such measures now exist in indices of productivity and biological diversity. Further development and deployment of EM will depend in part on the improvement and objective application of measures of ecological health. Another index of EM application and value will be better recognition and forecasting of future costs avoided and ancillary economic effects. It may be that regions and countries containing a high proportion of healthy ecosystems do better economically because, for example, they attract effective entrepreneurs and labor.

THE MANAGEMENT OF "NATURE"

It is increasingly clear that EM presents one of the only available ways to begin to resolve the paradoxical need to manage nature. EM contains the minimum set of activities that must in prudence be applied by people to any natural system, including the global one. What this "EM minimum set" instructs us to do with natural systems is again fivefold and includes instructions to (1) keep system parts, (2) keep accurate records, (3) make forecasts, (4) test them, and (5) look outward from, as well as specifically at, the system in question.

CONCLUSION

EM has a solid conceptual base and is being overtly applied on some public and private lands. It will be increasingly important to document these efforts so that retrospective study can result in improved concepts and methods. Even more important is the task of developing agreed-upon measures of ecosystem health and sustainability.

PART III

Social and Policy Considerations in Defining Sustainable Forestry

Introduction

V. Alaric Sample
Forest Policy Center, American Forests

Is the definition of sustainable forestry purely ecological, or is there a deeper dimension to the concept? The preceding chapters have explored some of the ways in which an ecosystem approach to managing forests at the landscape scale might differ from the current approaches to managing the major U.S. forest types. Public interest and concern for the health and well-being of the world's forests have never been greater, motivated largely by the unprecedented rate of loss of species and other biological issues. This concern, in turn, has stimulated a renaissance in forestry research aimed at improving our understanding of the complex functioning of natural forest ecosystems and how they respond to various kinds of natural and human interventions. Clearly, many scientific and technical questions remain to be answered. But many additional resources and some of the best minds in the natural sciences have been dedicated to answering these questions, and the rate of learning continues to accelerate.

When we have cleared the remaining scientific and technical hurdles, will we have achieved or even fully defined sustainable forestry? Once we have reached a common understanding of what changes in current forest management practices are implied in the evolution toward sustainable forestry and have developed the scientific understanding of ecosystem functioning and response necessary to achieve desired resource conditions, we must still confront a host of social, political, and economic barriers to the implementation of these changes. The determination of *what* we need to be doing differently from the way we have managed forests in the past must be accompanied by some notion of *how* we might bring about those changes. The organizations and institutions upon which society is structured, both within and outside the resource management

247

professions, are not well suited to make fundamental adaptations that keep up with the current rapid pace of change.

In this section, the authors explore some of the changes in our social, political, and economic institutions that will be necessary to facilitate a shift to sustainable forestry—which we now recognize will be defined not only ecologically but also by evolving social values.

Alice Rivlin takes a broad approach that challenges the adequacy of our current approach to political decision making and our ability to achieve sustainable forestry without some fundamental changes in that system. The ecologists and resource management specialists have given a clear and consistent message that we need to take a broader view of forests, both in the way we perceive them and in the way we manage them. There is broad agreement that we must move to a management system that gives greater weight to values other than wood production, that gives greater weight to the future over the present, and that preserves options until we understand better the functioning and response of complex biological systems. Rivlin asks, How do we go about this when our entire social and political system is weighted the other way? Our political system is one that gives low value to future benefits and to diffused social benefits that cannot be directly appropriated by the market and that is designed to resist change and the pain of change. Tolerating the existing situation seems less painful than figuring out how to manage change, and the forces opposed to change tend to capture the political system.

Rivlin sees three types of remedies that seem at first to have nothing to do with forests but, she asserts, are important to the kind of thinking that it will take to bring about the institutional changes needed to move us in the direction of sustainable forestry. First is a revolution in education about public choices to help people understand the consequences of the choices that they are making, whether by deciding or by not deciding. Second, we need policies to assist people and communities that are victims of change to adjust, to retrain, to move, and to have new ideas about what is going to support their community—rather than simply to resist change. Third, major campaign finance reform may be needed to move away from a political system in which those who stand to lose or gain in the short run have the lion's share of the electoral power.

The current and potential roles of economics in facilitating the shift toward sustainable forestry are examined from both the theoretical perspective, by Michael Toman, and the forest manager's perspective, by

William Ticknor. Ticknor observes that defining sustainable forestry is ultimately a question of understanding evolving social values and that the existing system forces forest managers to become arbiters of the conflict between private and community values, a position for which they are "ill-equipped by virtue of either temperament or training." Thus far, economics has been of little help in guiding managers through these troubled waters. Particularly from the perspective of private forest landowners, "economics acts as a disincentive to practice forest management consistent with our emerging social values." *Homo economicus* has little incentive to moderate the quest for wealth by other concerns. Our economic system and the functioning of the market are "frustrating the evolution of *Homo ecologicus.*" Ticknor calls for a recognition of the problem of *H. economicus* and the need to develop a business environment more supportive of the long-term health of the biosphere.

Toman describes a fundamental difference in the way ecologists and economists define *sustainability.* For ecologists sustainability connotes preservation of the status and function of ecological systems; for economists, the maintenance and improvement of human living standards. Disagreements about the salient elements of the concept hamper determination of appropriate responses for achieving sustainability. Key topics about which disagreement arises include intergenerational fairness, the substitutability of natural and other resources, and the carrying capacity of natural ecosystems. Disparate perspectives on these topics might be bridged through the concept of the safe minimum standard, which posits a socially determined demarcation between moral imperatives to preserve and enhance natural resource systems and the free play of resource tradeoffs.

Several authors have made contributions to the construction of a framework of social and political considerations that parallels the ecological elements of sustainable forestry. Like Rivlin, however, they see a far broader need for changes in these systems than merely their relevance to achieving sustainable forestry. The notion that nature is intrinsically good and that the natural balance would be restored if only humans would stop affecting the environment is rejected as barely relevant in a rapidly industrializing world of 5.5 billion people, with nearly twice that population expected within a century.

Ticknor calls for a new integrating principle for sound inputs to a more holistic decision process, one "capable of dealing with multiple dimensions—biological, physical, emotional, and spiritual—across geographic scales ranging from microscopic to regional to transnational and, ultimately, global." Presently, he notes, our whole institutional approach to problem solving, our "preoccupation with minutiae and reductionist tendencies make it difficult to focus adequate attention on holistic, systems approaches to the myriad problems that confront us." In a parallel to Jerry Franklin's assertion of the declining usefulness of the approach by both the environmental and development communities to protect biological diversity through a system of set-aside preserves, Ticknor observes that the polarized, adversarial approach to resource policy making has brought the process to a dead halt—just at a time when it is most important that the policy process be flexible, dynamic, and responsive to rapid changes in social values and the scientific understanding of forest ecosystems.

Jeff Romm takes a more indirect approach to the need to move beyond the adversarial model of resource management decision making, first drawing a distinction between sustainable forests and sustainable forestry. The definition of a *sustainable forest*, a forest's desired qualities, will vary depending upon what different people want and over how large an area and over what length of time they assess the gains and losses. Because people do not define forest sustainability in the same way, it becomes a political term, a term that connotes conflict. But Romm views such conflict positively, as a means of mobilizing diverse interests "in preserving, enhancing, using, or redistributing resources over different areas and times to challenge existing structures of social control, adaptation and change." *Sustainable forestry*, on the other hand, is viewed not as a set of forest conditions, but as a social process that maintains opportunities in the future to accommodate a diverse set of needs and values and "maintains a climate to encourage adequate further investment." The contemporary conflicts about forest sustainability, says Romm, "reflect the gap between the diversity of preferred 'sustainable forests' and the capacity of social processes to resolve differences for mutual benefit." The result is that conflict has reached seriously counterproductive levels.

The struggle between those who would log and those who would preserve the forest has taken on a life of its own, and it is a struggle in which the forest itself may ultimately lose. Fearing the uncertainty of the future regulatory climate, industry is reducing investment in forests while cutting as quickly as possible, hastening its own political doom. Those who would preserve the forests may find that, by diminishing the economic value of the nation's predominantly private forest lands, they have unwittingly hastened the replacement of forests with subdivisions. Like the global superpowers during the Cold War, the industry coalition has been kept together by the existence of the preservationist coalition, and vice versa. As others have observed, "without this bogeyman effect, each side would fly apart into its many constituent parts. . . . Local environmentalists in timber towns would split off from national environmentalists, and loggers and mill workers would no longer line up with timber firms."

In Romm's view, this would not be an altogether bad thing. There may be no single global or even nationwide definition of sustainable forestry, but instead a diversity of definitions based on a recognition of the unique characteristics of the local situation with respect to other considerations at larger geographic and political scales. Sustainable forestry is seen as a set of social processes that recognize and accommodate diverse and dynamic perspectives of what a forest should be. Romm describes the requisite constituents of an approach that will help "recreate a civil fabric within which people can at least agree to accept the possible legitimacy of others' views, to stem the hemorrhage of trust and faith in the future that any forms of investment require, to reverse the disinvestment that has put the industry, the workers, and the environment together on a steep-diving roller coaster . . . [to] either develop the means to work together or . . . lose our forests as any of us know and want them."

In her look at sustainable forestry's implications for economic growth and development, Julie Gorte puts a human face on such cool, abstract terms as *the free play of resource tradeoffs*, and the face is often one of uncertainty and anguish. Moving toward sustainable regional economies will, in some cases, require adjustments that disrupt current economic patterns, particularly in rural, forest resource–based regions and communities. Sustainability for communities hinges on their resilience and their ability to adapt to change, whether from internal or external forces. Trying to prevent change in the face of overwhelming economic forces is one sure way of guaranteeing economic disruption, if not now then in the future when some unexpected storm finally overwhelms the dike erected to stay the steadily rising tide.

The example of Lowell, Massachusetts, a former textile center now known for computer and electronics manufacturing, may hold some lessons for the Pacific Northwest as it looks for comparative advantages to encourage new economic enterprises. These might include enterprises radically different from those that have characterized the regional economy in the past. Large urban or regional economies are more adaptive than are small, local economies, and they need help to adjust to changed conditions. Resources exist in the form of federal and state programs to aid workers and communities. (The Job Training Partnership Act was recently tripled to deal with defense build-down and the effects of protecting the spotted owl. Displaced workers in California, Oregon, and Washington have on average found jobs at a higher rate and wage than have workers in other regions.) But first there must be a willingness to adapt and adjust to new and different economic pursuits, rather than a dogged resistance to change or a persistent denial that change is necessary.

After recognizing and defining some of the vital changes that must take place in organizations, institutions, and society to achieve sustainable forestry, we must take the specific actions that will bring about those changes. Robert Perschel makes this challenge a highly personal one because, ultimately, it cannot be otherwise. Organizations and institutions are collections of individuals. Individuals created them and endowed them with their vision and energy. For an organization not to become stagnant and irrelevant, it must be recreated each day and endowed with new energy by its constituent individuals. This daily process affords opportunities to keep the organization vibrant and adaptable so that individuals can lead their fellow constituents forward to the achievement of the highest goals of which the organization is capable.

Thus, Perschel argues, it is not enough to identify the changes that are needed in our institutions and then sit back and wait for them to change. The change process within any institution begins with single individuals who, through their personal leadership and persuasion, invoke the energies of other individuals to meet and overcome the challenges that confront them. Perschel sees Americans as having a special ability—and responsibility—to lead in making the changes in their institutions that are necessary to facilitate sustainable forestry: "Our Constitution offers the autonomy and freedom needed for the individual to achieve in accordance with the highest standards of the human spirit. . . . Freedom and autonomy are an American privilege, and with it comes the obligation to lead." More than in perhaps any other nation in the world, Americans have the freedom to make their institutions and organizations flexible and responsive to our new understanding of the importance of protecting and managing our forests in a way that is both economically and environmentally sustainable. And with that freedom comes the responsibility not only to manage our own forests sustainably, but also to assist other nations of the world in discovering and then using the keys to sustainably managing their forests, for themselves and for the global community.

It is clear that there are social, political, and economic—as well as ecological—components in any comprehensive definition of sustainable forestry. We even have a reasonably clear notion of what those components are and how they will have to change from their present states to move in the direction of sustainable forestry. As with the ecological considerations, however, we are providing far more new questions than answers. At this stage in our effort to define sustainable forestry, that is probably appropriate. Better to spend the time to be sure that we have formulated the right questions and posed them as completely and accurately as we can than to find later that we have come up with the precise, right answers to all the wrong questions.

10

Values, Institutions, and Sustainable Forestry

Alice M. Rivlin
The Brookings Institution

Thus far, the discussion of sustainable forestry has generally involved foresters, biologists, ecologists, and other scientists. I bring a different perspective—that of an economist who has spent most of her career immersed in the formulation and implementation of public policy at the national level. In this chapter, I will draw on my experience with politics and policy making to speculate about the changes in social values and institutional attitudes that would be necessary to facilitate evolution toward sustainable forestry in North America. That is a big task. I can only hope to throw out a few ideas that will start people thinking.

There exists no consensus on exactly what sustainable forestry is or what brings it about. The emergence of such a consensus, if it is even possible, will require persistent discussion through many more conferences, research papers, and debates. The thoughtful interchanges at these sessions, however, reveal some of the important elements of the evolving concept of sustainable forestry. One such element is the recognition that forests are enormously valuable to the earth and its inhabitants for a multiplicity of reasons, only one of which is wood production. Forests maintain water supplies and preserve air and water quality. They provide habitat for all manner of creatures and preserve much of the earth's biodiversity. They are rich environments of great beauty. They play a major role in climate.

A second element in the evolution toward sustainable forestry is the recognition that a forest is very different from a tree farm. A forest is a

255

much more complex system. Surprisingly little is known about how forests actually work, how they sustain themselves, how they evolve, or how they break down.

The level of human knowledge about forest systems is similar to the level of human knowledge about other complex systems such as market economies and urban communities. Failure to understand how complex systems work can lead to expensive, even irreversible mistakes. Putting up fancy new buildings in the center of decaying urban areas, such as Detroit, for example, does not produce a viable urban community. Communities are complex and need far more than buildings to sustain themselves. Similarly, sustainable forestry will not result from lengthening rotations on tree farms and preserving a few small areas for display of other forest qualities.

The evolution toward sustainable forestry requires, at a minimum, a recognition of the limitations of current knowledge and of the risk that human intervention will do irreversible harm before enough knowledge accumulates to identify the practices of sustainable forestry. This recognition leads to a double strategy: (1) intensify research on how forest systems work and (2) preserve options for the future. Preserving options implies stopping policies that are doing obvious harm by destroying watersheds, biological diversity, scenic beauty, and other forest values. It means developing new forest management techniques that give far less weight to the present and more to the future and less weight to wood production and more to other values. Slowing the destruction is necessary lest humans lose the forests before they really figure out how they can be sustained or what is being lost.

Even these first, tentative steps in the direction of sustainable forestry will be difficult because the whole structure of North American political and social institutions is aimed in the opposite direction. First, our political process tends to assign low weights to future as compared to present benefits. Voters tend to ask politicians, What have you done for me lately? They do not ask, What have you done to preserve options for my children and grandchildren? Elected officials, who must face the voters frequently, tend to choose quick results over long-term vision.

Second, the political decision process emphasizes benefits that can be identified, measured, appropriated, and valued in monetary terms—so many million board feet of timber sold, for example. The process tends to neglect diffused social benefits that cannot be easily valued or even measured, such as the benefits of clean air, wildlife habitat, or scenic beauty.

Moreover, the social and political system of North America, like most others, is designed to resist change. Those who are hurt by change tend to get more attention than do the beneficiaries. The political system is willing to subsidize the status quo (to subsidize timber-dependent communities, for example) because that is less painful than figuring out how to manage change. The forces opposed to policy change tend to have the resources and organization to capture the political system, while the beneficiaries of change do not.

I offer three remedies for the institutional and social biases that impede evolution toward sustainable forestry. These remedies do not seem at first to have anything to do with trees. Indeed, they would contribute as much to a movement toward sustainable economies or sustainable urban communities as toward sustainable forestry.

First, we need a revolution in education about public choices. We need to find ways of helping people to understand the consequences of the choices they are making, implicitly or explicitly, about the world in which they or their children will live. Daniel Yankelovitch, in a provocative book called *Coming to Public Judgment,* pointed out that we worry about public opinion a great deal in this country, but we do not put much effort into improving the *quality* of public opinion. We make concerted efforts to improve the quality of lots of things—food, schools, roads, water—but hardly any thought or energy goes into improving the quality of public opinion.

Yankelovitch defined a high quality of public opinion as a situation in which people understand the main consequences of the choices to be made and are willing to pay the price of the choice they favor. In many areas of public policy, there is evidence that high-quality public opinion does not exist. For example, polls show that most people oppose large federal deficits, but they are not facing up to the consequence of that choice. They also favor lower taxes and more public services, so they are not willing to pay the price of the deficit level they say they favor. The savings and loan debate is another example of the public and its elected representatives failing to think through the consequences of the choices they are making. By freeing thrift institutions from constraints while continuing to insure deposits, they created incentives for institutions to take excessive risk with depositors' money. Similarly, in many parts of the country, people talk earnestly and emotionally about preserving timbering as a way of life while supporting policies, such as clear-cutting old-growth forests, that will undermine that way of life in the future.

Hence, we need ways to help communities think about their futures and consider the consequences of the choices that they are making. Timber-dependent communities need to consider what kind of communities they want and the consequences of alternative ways of managing the forest for the future. Technology can help. Highly sophisticated mapping and computer modeling can be used to show graphically the consequences for the landscape and the water and the future timber yield of alternative forest management strategies. People can see not only the consequences of past choices but also the likely result of current ones. They can use the technology to ask "what if" questions and discuss the consequences for their community of alternative timber futures. One can now imagine town meetings in which people crowd around the computer screen to experiment with different options and talk about them. Do we really want to cut trees this fast? What will this cutting rate mean for jobs or water or air quality in five or ten years? What are the consequences for loggers, for sawmills, for hunting, for tourism, for new industry?

Second, we need society-wide efforts to manage change in constructive ways. There would be less resistance to change if we had programs ready to assist both individuals and communities that are victims of change. For individuals, assistance could involve retraining, help with the job search or relocation, or credit to help start a new business. For communities, assistance could be help in considering alternative futures and finding new economic bases for community prosperity.

Policies to manage change should not be specially focused on forest-dependent communities. Timber towns are only a fraction of the communities facing change. Individuals and communities all over the country are dealing with the result of rapid technological change, of new patterns of world trade and commerce, and of the current shift from the Cold War to a new and still evolving world order.

All of these changes have major benefits. Technological advances, defense budget reductions, and increases in world trade can lead to higher standards of living, but they also have victims. We need a new national determination to manage change so that its benefits are maximized and its victims are helped to adjust. At present, we pay high prices for resisting change and postponing adjustment. Steel quotas, below-cost timber sales, and the continued operation of unneeded military bases are all prices paid by society to retard change. A new national mind-set of fostering adjustment to change could make the transition to sustainable forestry easier.

Third, neither the transition to sustainable forestry nor the evolution of sustainable futures will be possible without campaign finance reform. At present, those who oppose change have disproportionate political power. Industries that might sustain short-run losses as a result of change have more political clout than the citizens and industries that are likely to gain in the long run. Timber and mining industries that stand to lose from environmental preservation, steel or auto industries that may lose from free trade, defense industries that suffer from lower military expenditures—all have political action committees and can mobilize more campaign contributions than the beneficiaries of these political shifts. If the United States is to manage change in its long-run interest—including a transition to sustainable forestry—it must do what other democracies have done. It must shift to public financing of campaigns, free air time for proponents as well as opponents of change, and more chances for people to consider the consequences of alternative futures.

The dream of a sustainable future can spread far beyond the current discussion of forest management to the whole nation. We could have a serious national debate (or a series of local ones) about alternative futures. We could have a political system that makes possible the actualization of policies that most people want and eases the transition to more sustainable futures—in forestry and other parts of North American life.

11

Sustainable Forestry: Redefining the Role of Forest Management

William D. Ticknor
W. D. Ticknor Forestry Consultants, Inc.

In his marvelously insightful history of human ideas, coincidentally titled *The Crooked Timber of Humanity*, Isaiah Berlin wrote that "the constant theme which runs through all Utopian thought . . . is that once upon a time there was a perfect state, then some enormous disaster took place . . . the pristine unity is shivered, and the rest of human history is . . . an agonized effort to piece together the broken fragments of the perfect whole with which the universe began, and to which it may yet return" (Berlin 1990).

"An agonized effort . . . to restore a pristine unity" accurately describes the present tensions in the institution of forestry; we are making an agonized effort to perform an impossible task. The profession has lost its bearings and is suffering from acute vertigo. The traditional guidance systems—economics, growth and yield, forest health—have been found wanting in the difficult terrain we are flying over. There is a chorus of confusing voices urging change upon us: higher, lower, faster, slower. The altimeter reads a comforting 20,000 feet, but we don't know if we are over Hawaii or approaching the Himalayas. We have identified no common destination. The only consensus that has emerged from the clamor is agreement that we cannot stay where we are.

Today's forest manager is not unlike the pilot of a mythical aircraft whose passengers have all agreed to leave point A but who have not agreed about the location of point B. So the pilot, in his wisdom, having elected to touch down at the average destination preference for everyone on board, has just filed a flight plan to ditch in the middle of Lake Erie. And he's happy to find that some of his passengers think that Lake Erie is a suitable place for a ditch. Like passengers on an allegorical airplane, we have decided to change the *way* we manage forest resources, but we have not decided *how*. In the meantime, we've laid an extraordinarily heavy burden on conscientious resource managers who are seeking to respond to the chorus of concerned voices, the pleas to treat innumerable legitimate issues as equally of first priority. And the passengers tend to blame the pilot for not knowing where they want to go.

As Bonnicksen (1991) pointed out, the task of forest managers is to mediate between the society that creates and validates their role and the resource that they are trained to manage. They are not autonomous agents who exist to benefit the resource; rather, they represent the interests of their employers in interactions with the resource. Basically, they work for us. As private individuals, we own the majority of the commercial timberland base in the United States. Collectively, we own the public lands and, directly or indirectly, we determine how they are to be managed. We own the corporations; we elect their directors; we work for them, manage them, support them in the marketplace, and regulate them. It may take a little while, but, sooner or later, corporations will do the will of the society in which they exist.

We are the ones who make the decisions about what resources we will use or abuse, what environmental damage we will countenance, and what value we assign to forest ecosystems and biodiversity. Malcolm Hunter (1990) put it this way: "As long as people are manipulating ecosystems, it is inevitable that their value systems will influence their goals." Ultimately, forest managers and the programs they implement reflect the demands and values of the society for which they work. Society defines the role of resource management.

If the forest managers of the future are to perform their task effectively, they must have the requisite technical and social skills. They must be comfortable and capable with both traditional forestry skills and a wide variety of new competencies, including ecosystem dynamics, the principles of forest landscape management, and the fundamentals of biological diversity, to name just a few.

Society also has its share of responsibilities in this scenario. We must clearly define the parameters of our social contract with forest resource managers, and we must elucidate the values that will guide their practice on our behalf. What are these voices presently saying? In the language of the marketplace, we are saying, "Increase the quantity and quality of economic consumer products and of jobs." In the language of steward-ship, we are saying, "Increase the quantity and quality of noneconomic public goods." We have not suggested to the manager how this duality can be accomplished.

Sorting through this confusion, three pivotal difficulties seem to emerge in our contract with practicing land managers: a sociopolitical problem, an economic problem, and a technical problem. The political problem is the easiest to describe. It will, perhaps, be the most difficult to resolve. The question is: How shall we make resource management decisions in the future? What role will the resource manager have in making these deter-minations? To whom and to what process will the manager look for guid-ance? Decision mechanisms that have served us reasonably well during the epoch of economic man are not functioning adequately as we deal with issues that extend beyond the traditional boundaries of space and time.

The second problem category is economics. To what extent should we expect resource owners to subsidize the sustained production of public goods? Do we look to resource managers to determine the appropriate balance between economic and noneconomic values?

The third is a technical problem. Once it has been determined which aspects of the forest ecosystem and its human components should be incorporated in a management program, what algorithm or integrating principle can the manager use to combine these aspects in practice, on the ground, in an optimum way?

I will discuss briefly each of these concerns in relation to sustainable forestry. As I use it, this term involves manipulating the forest biome to obtain valued amenities, using practices that maintain the variability, vitality, and resilience of forest ecosystems in a continuously changing global context. Crafting high-quality biospheric decisions represents a social exercise that is totally different from anything the old world order has ever attempted. In the era of economic man, the forest manager's primary source of decision data was embodied in the answer to one question: What does the owner want to do with it?

Hawley and Smith, in the 1954 edition of their well-respected text, *The Practice of Silviculture*, described the purpose of silviculture as "the production and maintenance of such a forest as will best fulfill the objec-tives of the owner and yield him the highest return in a given time." Generations of foresters took this definition very seriously—and still do.

In the new era, two different questions are emerging for the manager. First, how does the nature of the property itself, together with its landscape setting, suggest that the property should be managed? As this question intimates, to a significant degree the character of the land itself may predetermine the most appropriate management prescription. Second, to what extent should broad ecological and social considerations be integrated into the management program?

As these questions indicate, the volume and variety of information managers must deal with are changing radically. Managers must cope with two different but related categories of problem. One embraces questions for which there are or will be scientific answers, objective questions for which definitive quantitative answers will ultimately be found.

- To what extent does clear-cutting of Appalachian hardwood stands result in site deterioration, and by what quantity, if any, is sustainable productivity reduced?
- At what age and stocking level is the rate of carbon sequestration maximized in slash pine plantations?
- What amount of habitat variation can the red-cockaded woodpecker tolerate?

Sooner or later, our management decision process will be informed by reliable answers to these questions, but the answers, contrary to our wishes, will seldom be couched in terms of right or wrong, yes or no. They require the election of alternatives, the exercise of judgment, and the action of choosing. The Scientific Panel on Late-Successional Forest Ecosystems in its draft report to the U.S. House of Representatives made this point eloquently in a concluding paragraph: "We think that we have prescribed a sound basis for decisions given the time and information limits within which we operated. Science has done what it can. The process of democracy must go forward from here." (Johnson et al. 1991).

That is a worrisome prospect. The question is: In what manner is democracy to move forward? The reason it's worrisome is that we have not developed an appropriate decision process. Surely, there must be a more constructive process than litigation. If objective issues are difficult to deal with, purely subjective issues are infinitely more confounding.

- To what extent shall we modify commercial timber practices so as to preserve aesthetic amenities?

- How will we balance the demands for development and green space—the preferences for clustered concentrations of human activity, on the one hand, or dispersed populations, on the other?
- If the only otherwise logical landfill site in a given locality includes the habitat of a globally imperiled prairie grass, and the grass is in my backyard, can we agree to site the landfill in your backyard?

I think we have learned the painful lesson that "not to decide is to decide." Now we must decide *how* to decide. From the point of view of forest management, the forest manager, and sustainable forestry, this a critical issue. Decisions must be made, and they must be communicated to the forest manager. We must devote as much creative attention to devising a quality decision process as to the quality of decisions themselves. As the World Commission on Environment and Development (1991) observed, changes in the scale of human activity and the rate at which change is occurring are "frustrating the attempts of political and economic institutions, which evolved in a different, more fragmented world, to adapt and cope." Forest managers are enmeshed in the frustration of which the commission speaks.

The second issue I want to address is the area of economics and the growing disparity between economic values and environmental values, between public goods and private goods. When I acquired my knowledge of forestry as a student and during thirty-five years of forestry practice, the primary measure of forest productivity was economic. Although other elements played a token role in the decision process, economic analysis was always recognized as the single most useful tool to discriminate between multiple management alternatives. I remember a terse synopsis of the philosophy, which I often shared with starry-eyed newcomers to the practice of industrial forestry: "Cash drives the system."

This is no longer the case. During the last decade, the value of non-economic forest outputs, which is to say the value of public goods, has escalated at rates greatly in excess of the rate of increase in the values of economic outputs. As one result of this disequilibrium, economic measures can no longer be considered reliable primary guides for forest-management decision making. Indeed, quite to the contrary, economic and environmental considerations often conflict, with the unfortunate and counterproductive result that economics often acts as a disincentive to practice forest management consistent with our emerging social values.

We should not be surprised by the intractability of this problem. After all, we have all been immersed in an economic view of life for a long time. As Daly and Cobb (1989) pointed out, "modern economic theory has taught that . . . inhibitions of the quest for wealth are not needed for the general good and that, indeed, they impede its realization. Where each individual seeks to maximize economic gain, the total product of society increases and hence all people benefit. . . . *Homo economicus* has little incentive to moderate the quest for wealth by other concerns." The forest manager becomes the arbiter of the conflict between private and community values. It is a position for which the manager is ill-equipped by virtue of either temperament or training. As we struggle with impediments to sustainability, we will want to recognize the problem of *Homo economicus* and develop a business environment more supportive of the long-range health of the biosphere.

We can overcome economic disincentives with regulations and penalties and taxes, and this seems to be the route we prefer. But I suggest that it will be far more effective to obtain alignment of economics with environmental values, so that economics reinforces good environmental practices. The considerable creativity and energy of forest managers will then more often be directed toward implementing beneficial programs instead of minimizing the costs of compliance.

If we understand anything about human motivation, we will recognize that we can accomplish far more with positive reinforcement than with punishment or a list of "do nots." Negative reinforcement tends to reduce instances of unacceptable behaviors, but we are interested in finding ways to increase the number of acceptable behaviors. Who feels better about a football game? The team that won, or the team that had the fewer penalties? Who would we expect to do a better job of management? The individual who is rewarded and recognized for doing right? Or the person punished for doing it wrong?

As we participate in and promote the critical evolution from *Homo economicus* to *Homo ecologicus*, let us recognize that the economic environment that well served the interests of economic man is, in many respects, inimical to the well-being, perhaps even the survival, of ecological man. If our social contract with resource managers is to be optimally effective, we will want to create an economic climate supportive of our mutual interests.

I turn, lastly, to the technical dimension of sustainable forestry, which, again, I want to address from the perspective of the forest resource manager. We are moving forward very rapidly in the process of identifying critical components and systems of the forest ecosystem—the parts and how they are related to one another.

My sense is that we are doing less well at understanding the global forest ecosystems from a holistic perspective. We are desperately in need of an integrating principle, a model that can provide sound inputs to the decision processes we face. This decision engine must be capable of dealing in multiple dimensions—biological, physical, emotional, spiritual—across geographic scales ranging from microscopic to regional to transnational and, ultimately, global. This decision algorithm must recognize the legitimate private ownership prerogatives that remain after all of the mandatory public interests, present and future, have been served. The model must enable us to move toward a new way of assigning priorities to the individual elements, the parts and subsystems, of the global landscape and to the entire biotic community, absolutely including the human species.

Think of a car; what's more important? The fuel? The starter? The brakes? None of the above, of course. The *system* is important, and a malfunction of any component becomes important because it can cripple the entire system. The appropriate model for sustainable forestry will help us skillfully combine programs for

- commodity production,
- aesthetics,
- biodiversity conservation,
- CO_2 sequestration,
- energy conservation,
- site productivity, etc.,

all in an optimally harmonious way. The model will help us move beyond our preoccupation with "either/or" and "more or less" to a "both/and" view, not seeing social values as more or less important than commodity values but recognizing that the sustainable quality of global life depends on both.

I have observed before, but it bears repeating: the successful forest management of the future will be a right-brain enterprise with a systems orientation (Ticknor 1990). Traditional forest science is the essence of a left-brain, reductionist endeavor—analytical, quantitative, logical, linear. It is a science in which the whole is understood in terms of its elements. If anything symbolizes reductionist forestry, it is the sample plot. Most of us who are practitioners of this science enjoy and excel at left-brain pursuits, and our proclivities in this direction were reinforced by several years of training.

But the decision process we must move toward will be a more intuitive, combinatorial approach that will place a high priority on blending the scenic, aesthetic, and spiritual aspects of forestry with the biological and business aspects. One of the best predicates for this new approach is architecture, in which a great deal of attention is devoted to the artistic blending of form and function, attaining excellence in providing practical uses of space artistically arranged to please our aesthetic senses. It is an interesting commentary on our value system that many of the corporations we invent and own and manage spend millions to make architectural statements but are unwilling to depart from shrewd, pecuniary financial analysis in managing forest resources.

The future for right hemispheres in resource management is bright indeed. Computers are taking over the left-brain tasks and providing marvelous tools for use in simulating the spatial and integrative components. We now need to place these tools in the hands of individuals not hampered by the confining restraints of convention. And we need to balance our propensity to focus on increasingly smaller scales of activity with a systems view. Peter Senge (1990) provided this succinct description of systems thought:

Systems thinking is a discipline for seeing wholes. It is a framework for seeing relationships rather than things, for seeing patterns of change rather than static "snapshots." . . . And systems thinking is a sensibility—for the subtle interconnectedness that gives living systems their unique character.

Today, systems thinking is needed more than ever because we are becoming overwhelmed by complexity. . . . The scale of complexity is without precedent.

Complexity can easily undermine confidence and responsibility. . . . Systems thinking is the antidote to this sense of helplessness. . . . [It] is a discipline for seeing the "structures" that underlie complex situations. . . . Without systems thinking there is neither the incentive nor the means to integrate the learning disciplines once they have come into practice.

Our science of the environment and of the global forest resource is expanding rapidly. We are obtaining important insights at rates that greatly exceed our information-processing and decision capabilities. Our economic system and the artifacts and conventions of economic man are frustrating the evolution of *Homo ecologicus*. And our institutional preoccupation with minutiae, our reductionist tendencies, make it difficult to focus adequate attention on holistic, systems approaches to the myriad problems that confront us. It is critical that we devote actively creative energies to the sociopolitical, economic, and systems aspects of the sustainable global forest management task.

To this end, Rosabeth Moss Kanter, Professor of Sociology and of Organization and Management at Yale, wrote that "the last critical force for guiding productive change involves making sure there are mechanisms that allow the new action possibilities to be expressed. The actions implied by the changes cannot reside on the level of ideas, as abstractions, but must be concretized in actual procedures or structures or communications channels or work methods or rewards" (Kanter 1983).

In conclusion, then, I ask, Do we have the mechanisms in place which will, as Kanter puts it, "allow the new action possibilities to be expressed"? Clearly, we do not. It is, however, well within our scope to initiate a process to get those needed mechanisms in place. It is tempting to assume that someone else will take on that responsibility. If they do, fine. But if they do not, what then? Shall we let future generations wonder how it could have been that we saw the opportunity for action but failed to seize it? I, for one, certainly hope not.

REFERENCES

Berlin, I. 1991. The crooked timber of humanity. In *The history of ideas*. New York: Alfred A. Knopf.

Bonnicksen, T. M. 1991. Managing biosocial systems: A framework to organize society-environment relationships. *Journal of Forestry* 89:10-15.

Daly, H. E., and J. B. Cobb, Jr. 1989. *For the common good: Redirecting the economy toward community, the environment, and a sustainable future*. Boston: Beacon Press.

Hawley, R. C., and D. M. Smith. 1954. *The practice of silviculture*. New York: John Wiley & Sons.

Hunter, M. L. 1990. *Wildlife, forests, and forestry: Principles of managing forests for biological diversity.* Englewood Cliffs, NJ: Prentice-Hall.

Johnson, K. N., J. F. Franklin, J. W. Thomas, and J. Gordon. 1991. Alternatives for management of late successional forests of the Pacific Northwest: A report to the U.S. House of Representatives, Committee on Agriculture, Subcommittee on Forests, Family Farms, and Energy, and the Committee on Merchant Marine and Fisheries, Subcommittee on Fisheries and Wildlife Conservation and the Environment. Draft Report.

Kanter, R. M. 1983. *The change masters: Innovation for productivity in the North American corporation.* New York: Simon & Schuster.

Senge, P. M. 1990. *The fifth discipline: The art and practice of the learning organization.* New York: Doubleday/Currency.

Ticknor, W. D. 1990. Practicing objective forestry in a subjective world. Keynote speech to the American Forest Council Future of Forestry Conference, Washington, DC, October 16.

World Commission on Environment and Development. 1987. *Our common future.* Oxford: Oxford University Press.

12

Defining an Economics of Sustainable Forestry: General Concepts

Michael A. Toman
Resources for the Future

Sustainability has become a new watchword for assessing human influences on the natural environment and resource base. A concern that economic development, the exploitation of natural resources, and the infringement on environmental resources are not sustainable is expressed more and more frequently in analytical studies, conferences, and policy debates. This is no less true for forest ecosystems than for other elements of the natural endowment.

To identify what may be required to achieve sustainability in the use of forests or other natural resources, one must have a clear understanding of what sustainability means. Traditionally, the term *sustainability* has referred simply to a harvesting regimen that could be maintained over time. That meaning has been considerably broadened by ecologists to express concerns about preserving the status and function of entire ecological systems. From this point of view, sustainable forestry requires the perpetuation of the whole range of forest species and their relationships to each other, not just harvest capacity. Economists, on the other hand, usually have emphasized the maintenance and improvement of human living standards, in which natural resources and the environment may be important but represent only part of the story.

Beyond ambiguity of meaning there also is disagreement about the prospects for achieving sustainability. Some scholars essentially question whether sustainability is a significant issue, pointing out that humankind consistently has managed in the past to avoid the specter of Malthusian scarcity through resource substitution and technical ingenuity (Simon 1981). Others believe that the scale of human pressure on natural systems already is well past a sustainable level (Daly and Cobb 1989; Ehrlich and Ehrlich 1990). They point out that the world's human population probably will at least double before stabilizing and that, to achieve any semblance of a decent living standard for the majority of people, the current level of world economic activity must grow, perhaps fivefold to tenfold. They cannot conceive of already stressed ecological systems tolerating the intense flows of materials use and waste discharge that presumably would be required to accomplish this growth.

Ascertaining more clearly where the facts lie in this debate and determining appropriate response strategies are difficult problems—perhaps among the most difficult faced by all who are concerned with human advance and sound natural resource management. Progress on these fronts is hampered by continued disagreements about basic concepts. To narrow the gaps, we must first identify salient elements of the sustainability concept, about which there are contrasts in view between economists and resource planners on the one hand and ecologists and environmental ethicists on the other.

KEY CONCEPTUAL ISSUES

Two elements seem to be essential in understanding sustainability. The first is intergenerational equity and the responsibility of the current generation to their descendants. The second is the degree of substitutability between natural resources, including the environment, and other forms of social capital. Most disagreements about the concept of sustainability seem to reduce to disagreements over these basic points. In this section I review alternative perspectives on both issues, beginning with intergenerational equity.

The standard approach to intergenerational tradeoffs in economics involves assigning benefits and costs according to some representative set of individual preferences and discounting costs and benefits accruing to future generations just as future receipts and burdens experienced by members of the current generation are discounted. One justification for discounting over time is that people are impatient—they prefer current benefits over future benefits (and weight current costs more heavily than future costs). Receipts in the future also are less valuable than current receipts from the standpoint of the current decision maker because current receipts can be invested to increase capital and future income (Lind 1982). Discounting also can be justified, even with a concern for intergenerational fairness, if real levels of well-being are expected to grow over time and it is believed that the marginal value of additional consumption (including nonmarket benefits) is inversely related to real income. In this case, greater weight should be put on the relatively poorer earlier generations.

Critics of the standard approach maintain that invoking impatience entails the exercise of the current generation's influence over future generations in ways that are ethically questionable (Kneese and Schulze 1985; Norton 1982, 1984, 1989; Parfit 1983b; Page 1977, 1983, 1988). The capital and growth arguments for intergenerational discounting also are suspect, critics argue, because in many cases the environmental resources at issue—for example, the capacity of the atmosphere to absorb greenhouse gases or the extent of biological diversity in forest ecosystems—are seen to be inherently limited in supply.

These criticisms do not imply that discounting should be abolished (especially since this could increase current exploitation of natural and environmental capital), but they do suggest that discounting might best be applied in tandem with safeguards on the integrity of key resources like ecological life-support systems. We return to this point in the next section. Critics also question whether the preferences of an "average" member of the current generation should be the sole or even primary guide to intergenerational resource tradeoffs, particularly if some resource uses threaten the future well-being of the entire species but are experienced only dimly by current individuals. Adherents of "deep ecology" even take issue with putting human values at the center of the debate, arguing instead that other elements of the global ecological system have equal moral claims to be sustained. Even if one accepts that human values should occupy center stage, it is difficult to gauge what the values held by future generations might be in comparison to currently held values (Parfit 1983a).

A second key component of sustainability involves the specification of what is to be sustained. If one accepts that there is some collective responsibility of stewardship owed to future generations, what kind of social capital must be intergenerationally transferred to meet that obligation? One view, to which many economists would be inclined, is that all resources—the natural endowment, physical capital, human knowledge and abilities—are relatively fungible sources of well-being. Thus, large-scale damages to ecosystems, such as degradation of environmental quality, loss of species diversity, widespread deforestation, or global warming, are not intrinsically unacceptable from this point of view; the question is whether compensatory investments for future generations in other forms of capital are possible and are undertaken (Hartwick 1977; Baumol 1986; Solow 1986; Dasgupta and Maler 1991). Investments in human knowledge, technique, and social organization are especially pertinent in evaluating these issues.

An alternative view, embraced by many ecologists and some economists, is that such compensatory investments often are infeasible as well as ethically indefensible. Physical laws are seen as limiting the extent to which other resources can be substituted for ecological degradation (Kneese, Ayres, and d'Arge 1971; Ayres and Miller 1980; Perrings 1986; Anderson 1987; Gross and Veendorp 1990). Healthy ecosystems, including those that provide genetic diversity in relatively unmanaged environments, can offer resilience against unexpected changes and the preservation of options for future generations. For natural life-support systems, no practical substitutes are possible, and degradation may be irreversible. In such cases (and perhaps in others as well), compensation cannot be meaningfully specified. In addition, in this view environmental quality may complement capital growth as a source of economic progress, particularly for poorer countries (Pearce, Barbier, and Markandya 1990). Such complementarity also would limit the substitution of capital accumulation for natural degradation.

In considering resource substitutability, economists and ecologists often differ on the appropriate level of geographical scale. On the one hand, opportunities for resource tradeoffs generally are greater at the level of the nation or the globe than at the level of the individual community or regional ecosystem. On the other hand, a concern only with aggregates overlooks unique attributes of particular ecosystems or local constraints on resource substitution and systemic adaptation.

The issue of substitutability also is central in the debate over the scale of human influence relative to global carrying capacity. As already noted, there is sharp disagreement on this issue. As a crude caricature, it is generally true that economists are less inclined than ecologists to see this as a serious problem, putting more faith in the capacities of resource substitution (including substitution of knowledge for materials) and technical innovation to ameliorate scarcity. Rather than viewing it as an immutable constraint, economists regard carrying capacity as endogenous and dynamic.

THE SAFE MINIMUM STANDARD

Concerns over intergenerational fairness, resource constraints, and human scale provide a rationale for some form of intergenerational social contract (although such a device can function only as a "thought experiment" for developing our own moral precepts, since members of future and preceding generations cannot actually be parties to a contract). One way to give shape to such a contract is to apply the concept of a *safe minimum standard*, an idea that has been advanced (sometimes with another nomenclature) by a number of economists, ecologists, philosophers, and other scholars (Bishop 1978; Norton 1982, 1991; Page 1983, 1991; Randall 1986; Weiss 1989).

To simplify, suppose that damages to some natural system or systems can be entirely characterized by the size of their expected cost and degree of irreversibility. These two attributes of damages are treated separately. The magnitude of cost can be interpreted in terms of opportunity cost by economists or as a physical measure of ecosystem performance by ecologists.

Irreversibility reflects uncertainty about system performance and the resulting human consequences. At one extreme, very large and irreversible effects may threaten the function of an entire ecosystem. At a global level, the threat could be to the cultural if not the physical survival of the human species. At the other extreme, small and readily reversible effects are relatively easily mediated by private market transactions or by corrective government policies based on comparisons of benefits and costs.

Economists are accustomed to valuing consequences of irreversibility in an uncertain setting (Krutilla and Fisher 1985). One can then legitimately ask whether expected cost and irreversibility need to be considered separately or whether the expected cost of irreversibility cannot simply be included in an overall measure of expected cost. There are two reasons why it may be appropriate to treat cost and irreversibility separately. Monetizing all irreversibility suggests that compensating investment for any environmental degradation experienced by future generations is feasible and ethical. This is a debatable proposition, particularly among noneconomists. In addition, economic valuation inherently is individualistic. Keeping expected cost and irreversibility as separate criteria allows us to keep open the possibility of considering some values that are not purely individualistic, as advocated by some philosophers.

The safe minimum standard posits a socially determined, albeit "fuzzy," dividing line between moral imperatives to preserve and enhance natural resource systems and the free play of resource tradeoffs. To satisfy the intergenerational social contract, the current generation would limit in advance actions that could result in natural effects beyond a certain threshold of cost and irreversibility. Rather than depending only on a comparison of expected benefits and costs from increased pressure on the natural system, such proscriptions would reflect society's value judgment that the cost of risking these effects is too large. Possible resources for which society would not risk damages beyond a certain cost and degree of irreversibility include wetlands, old-growth forests, and other sources of genetic diversity.

There is a distinct difference between the safe minimum standard approach and the standard prescriptions of environmental economics, which involve getting accurate valuations of resources in benefit-cost assessments and using economic incentives to achieve efficient allocations of resources given these valuations (Bishop 1979; Smith and Krutilla 1979). Whether a resource-protection criterion is established by imperatives through application of the safe minimum standard concept or by tradeoffs through cost-benefit analyses, that criterion can be achieved cost-effectively by using economic incentives. However, for effects on the natural environment that are uncertain but may be large and irreversible, the safe minimum standard posits an alternative to comparisons of economic benefits and costs for developing resource-protection criteria. It places greater emphasis on potential damages to the natural system than on the sacrifices experienced from curbing ecological effects. The latter are seen as likely to be smaller and more readily reversible. In addition, the safe minimum standard invokes a wider, possibly less individualistic set of values in assessing effects. Since societal value judgments determine the level of safeguards, public decision making and the formation of social values are explicit parts of the safe minimum standard approach.

This illustrative discussion, of course, provides no actual guidance on where and how (if at all) such a dividing line between imperatives and tradeoffs should be drawn. There is uncertainty about how rapidly the threat to current and future human welfare might increase as damages become costlier and irreversibility becomes more likely. The location of the line will depend on the range of individual beliefs in society and available knowledge about human effects on ecosystems. For example, ecologists who are concerned mainly about irreversibility and believe that ecological systems are fragile might draw an essentially vertical line, with a large area covered by moral imperatives for ecosystem protection; economists who are concerned mainly about expected cost and believe that the well-being of future generations should be highly discounted might draw an essentially horizontal line, with little (or no) scope for moral imperatives. Acquisition of additional knowledge also will alter the relative weight given to imperatives and tradeoffs for specific ecosystems or the environment as a whole. In addition, how the delineation would be made depends on complex social decision processes, some of which probably have not yet been constructed.

The safe minimum standard thus does not provide an instant common rallying point for resolving the disagreements discussed here. However, this concept does seem to provide a frame of reference and a vocabulary for productive discussion of such disagreements. Such discussion would refine our understanding of what sustainability means and of the steps that should be taken to enhance prospects for achieving it.

CONCLUSION

Sustainable forestry ultimately is intimately wrapped up with human values and institutions, not just ecological functions. An entirely ecological definition of sustainability for forest management is inadequate; guidance for social decision making also is required. It must be recognized that human behavior and social decision processes are complex, just as ecological processes are. At the same time, economic analysis without adequate ecological underpinnings also can be misleading. What may seem self-evident to the student of the natural environment need not seem so for the student of human society, and vice versa.

The tension between ecological and economic perspectives on sustainability suggests several ways in which both economists and ecologists could adapt their research emphases and methods to make the best use of interdisciplinary contributions. Thus, for ecologists the challenges include providing information on ecological conditions in a form that could be used in economic valuation. Ecologists also must recognize the importance of human behavior, particularly behavior in response to economic incentives. Economists for their part could expand analyses of resource values to consider the function and value of ecological systems as a whole, making greater use of ecological information in the process. Economic theory and practice also could be extended to consider more fully the implications of physical resource limits that often are not reflected in more stylized economic constructs. In addition, research by economists and other social scientists (psychologists and anthropologists) could help to improve understanding of how future generations might value different attributes of natural environments. The sustainability debate should remind economists to distinguish carefully between efficient allocations of resources—the standard focus of economic theory—and socially optimal allocations, which may include intergenerational (as well as intragenerational) equity concerns.

ACKNOWLEDGMENTS

This chapter is adapted from Toman (1992). It reflects ideas developed jointly with Pierre Crosson (Resources for the Future) and Bryan Norton (Georgia Institute of Technology), whose help is gratefully acknowledged. This chapter also has benefited from helpful discussions with Al Sample (Forest Policy Center). Responsibility for the views expressed herein is the author's alone.

REFERENCES

Anderson, C. L. 1987. The production process: Inputs and wastes. *Journal of Environmental Economics and Management* 14(1):1-12.

Ayres, R., and S. Miller. 1980. The role of technological change. *Journal of Environmental Economics and Management* 7(4):353-371.

Baumol, W. J. 1986. On the possibility of continuing expansion of finite resources. *Kyklos* 39(2)167-179.

Bishop, R. D. 1978. Endangered species and uncertainty: The economics of the safe minimum standard. *American Journal of Agricultural Economics* 60(1):10-18.

————. 1979. Endangered species, irreversibility and uncertainty: A reply. *American Journal of Agricultural Economics* 61(2):376-379.

Social and Policy Considerations in Defining Sustainable Forestry

Callahan, D. 1971. What obligations do we have to future generations? *American Ecclesiastical Review* 144:265-280.

Daly, H., and J. Cobb. 1989. *For the common good.* Boston: Beacon Press.

Dasgupta, P., and K. G. Maler. 1991. The environment and emerging development issues. In *Proceedings of the World Bank Annual Conference on Development Economics 1990.* Washington, DC: World Bank.

Ehrlich, P. R., and A. H. Ehrlich. 1990. *The population explosion.* New York: Simon & Schuster.

Gross, L. S., and E. C. H. Veendorp. 1990. Growth with exhaustible resources and a materials-balance production function. *Natural Resource Modeling* 4(1):77-94.

Hartwick, J. M. 1977. Intergenerational equity and the investing of rents from exhaustible resources. *American Economic Review* 67(5):972-974.

Kneese, A., R. Ayres, and R. d'Arge. 1971. *Economics and the environment.* Baltimore: Johns Hopkins University Press for Resources for the Future.

Kneese, A. V., and W. D. Schulze. 1985. Ethics and environmental economics. In *Handbook of natural resource and energy economics,* eds. A. V. Kneese and J. L. Sweeney, vol. 1. Amsterdam: North-Holland.

Krutilla, J. V., and A. C. Fisher. 1985. *The economics of natural environments.* Washington, DC: Resources for the Future.

Lind, R. C., ed. 1982. *Discounting for time and risk in energy policy.* Washington, DC: Resources for the Future.

Norton, B. G. 1982. Environmental ethics and the rights of future generations. *Environmental Ethics* 4(4):319-330.

————. 1984. Environmental ethics and weak anthropocentrism. *Environmental Ethics* 6 (Summer):131-148.

————. 1989. Intergenerational equity and environmental decisions: A model using Rawls' veil of ignorance. *Ecological Economics* 1:137-159.

————. 1991. *Toward unity among environmentalists.* New York: Oxford University Press.

Page, T. 1977. *Conservation and economic efficiency.* Baltimore: Johns Hopkins University Press for Resources for the Future.

————. 1983. Intergenerational justice as opportunity. In *Energy and the future,* eds. D. MacLean and P. G. Brown. Totowa, NJ: Rowman and Littlefield.

————. 1988. Intergenerational equity and the social rate of discount. In *Environmental resource and applied welfare economics,* ed. V. Kerry Smith. Washington, DC: Resources for the Future.

————. 1991. Sustainability and the problem of valuation. In *Ecological economics: The science and management of sustainability,* ed. R. Costanza. New York: Columbia University Press.

Parfit, D. 1983a. Energy policy and the further future: The identity problem. In

Energy and the future, eds. D. MacLean and P. G. Brown. Totowa, NJ: Rowman and Littlefield.

————. 1983b. Energy policy and the further future: The social discount rate. In *Energy and the future,* eds. D. MacLean and P. G. Brown, 31-37. Totowa, NJ: Rowman and Littlefield.

Passmore, J. 1974. *Man's responsibility for nature.* New York: Scribners.

Pearce, D., E. Barbier, and A. Markandya. 1990. *Sustainable development: Economics and environment in the Third World.* London: Earthscan.

Perrings, C. 1986. Conservation of mass and instability in a dynamic economy-environment system. *Journal of Environmental Economics and Management* 13(3):199-211.

Randall, A. 1986. Human preferences, economics, and the preservation of species. In *The preservation of species,* ed. B. G. Norton, 79-109. Princeton: Princeton University Press.

Simon, J. 1981. *The ultimate resource.* Princeton: Princeton University Press.

Smith, V. K., and J. V. Krutilla. 1979. Endangered species, irreversibilities, and uncertainty: A comment. *American Journal of Agricultural Economics* 61(2):371-375.

Solow, R. M. 1986. On the intergenerational allocation of natural resources. *Scandinavian Journal of Economics* 88(1):141-149.

Toman, M. A. 1992. The difficulty in defining sustainability. In *Global development and the environment: Perspectives on sustainability,* ed. J. Darmstadter, 15-24. Washington, DC: Resources for the Future.

Weiss, E. B. 1989. *In fairness to future generations.* Dobbs Ferry, NY: Transnational Publishers.

13

Sustainable Forestry, an Adaptive Social Process

Jeff Romm

Department of Forestry and Resource Management, University of California, Berkeley

Defining Sustainable Forestry suggests a desire to resolve controversy about the nature of the sustainable forest. This implies the potential existence of some universally desirable state for the forest. However, this approach denies differences in human values that will not and perhaps should not go away. A forest's desired qualities, the defining virtues of the "sustainable forest," vary tremendously among people. Definitions depend upon what different people want, over what area and length of time they assess gains and losses, by what means they prefer to control the balance, and by and for whom they wish to exercise these means. *Sustainable forest* has no definition until the what-where-when-how-who, the value perspective, is specified. As few agree on these matters, the sustainable forest is controversial for good reason: any one definition represents particular values at others' expense.

The sustainable forest is an issue precisely because people cannot, and probably never will, define it in the same way. *Sustainable* is a political word that denotes conflict. It symbolizes resistance to changes people do not want and demands for changes that they do. It identifies a search for means to stabilize conditions and a social process that is bound to change them. It mobilizes diverse interests—in preserving, enhancing, using, or redistributing resources over different areas and times—to challenge existing structures of social control, adaptation, and change. It aggregates power among interests who, although eventually realizing that they differ, meanwhile can forge institutional realignments that modify the definition, valuation, and treatment of natural resources. In sum, it redefines the practical what-where-when-how-who, the systems of value, upon which existing structures are based.

Sustainable forestry is a different matter. Forestry is the regime of actions by which people conserve, augment, modify, and replace features of the forest so as to perpetuate its desired qualities, whatever these may be. It is a social process rather than a forest condition. Sustainable forestry augments this regime with processes for rapid adaptation to changes in what people need, want, and can do. Sustainable forestry is an adaptive social process that creates sufficient future forest opportunity to satisfy potentially competitive interests that would diminish the forest if left unresolved.

Contemporary conflicts about forest sustainability reflect the gap between the diversity of preferred sustainable forests and the capacity of forestry processes to resolve differences for mutual benefit. As a consequence, no one version of the good forest gains sufficient social support and investment to have a chance. In many parts of the world, conflict has reached seriously counterproductive levels. Conflict is eroding the very conditions under which support and investment on behalf of anyone's preferred forest become possible. Everyone's interests are going down together.

In California, for example, we face the urgent need to re-create a civil fabric within which people can at least agree to accept the possible legitimacy of others' views, to stem the hemorrhage of trust in others and faith in the future that any forms of support and investment require, to reverse the drain of forests that has put the industry, the workers and communities, and the environment together on a steep-diving roller coaster. We either will develop means to work together or we together will lose our forests as any of us know and want them.

The challenge is not to define *the* sustainable forest but to develop social processes that recognize, accommodate, and respond more effectively to diverse and dynamic perspectives of what the forest is and should be. The challenge is to achieve sustainable forestry.

SUSTAINABLE FORESTRY AS SOCIAL PROCESS

What social processes characterize sustainable forestry? The following seem to be useful categories:

- protection of people from external forces that discourage or prevent them from improving forest qualities,
- trade and cooperation among diverse forest interests,
- complementation among different levels and scales of forest governance,
- investment for future forest opportunities, and
- innovation.

Our search for sustainable forestry indicates the fundamental need to strengthen these processes. The search for means to do so is much more promising than a quest for the true sustainable forest and the static singular values that would define it. I discuss and illustrate these points with examples from Nepal, California, and India.

MANAGING EXTERNAL FORCES: THE FUELWOOD GARDEN AND THE REDWOOD FOREST

There is a huge tension between a norm of forest stability and the relentless dynamism of external forces—economic, political, environmental—on the people who shape forest conditions. The qualities of a forest depend upon how, when, and where external pressures affect the people who change or maintain these qualities. Few forest boundaries—the lines people draw to define the what-where-when-how-who that they want to include and exclude in a place—are strong enough to withstand these dynamic influences. Managing the forest is as much a problem of regulating external pressures as of responding to these pressures within the boundaries. Managing the forest to maintain some specific condition requires boundaries that adapt rapidly to contextual whims.

In a nursery in central Nepal, we once sought to see what would happen if a bed of eucalyptus seedlings, planted at a spacing of 1 inch by 1 inch, was let to grow rather than lifted for plantation. The soil had been cultivated deeply and was kept richly manured. Water supply and pest control were excellent. Skilled labor always was at hand. After several years, there had been almost no mortality. The bed looked like a virtually solid block of wood six feet tall. Harvesting a third of the stems and about half of the volume each year would satisfy a household's needs for cooking fuel on $^1/_{15}$ to $^1/_{20}$ acre of ground. If input supplies remained unlimited, the miniature forest could be sustained. If households used the system, their fuelwood-harvesting pressure on surrounding forest presumably would lighten, helping to conserve the forest for longer term needs (Romm and Seckler 1979).

But there is a large difference between an experimental plot and a farm household in which such a garden would need to work. A fuelwood garden would be just one part of a household economy, and its sustainability would depend on its costs to other parts of the farm operation. Household labor is scarce and would need to be taken from other activities, such as the production of food. Manure and water in such amounts would be drawn from a range that diminished opportunities for other households and villages needing these resources, with consequent costs of conflict or compensation. The worth of absorbing these costs would depend upon the relative ease or difficulty of getting "free" firewood from surrounding public forests. The requirements for sustainability in a farm are more demanding than those that apply in a well-protected experimental plot.

The problem does not stop at the farm level. If a household created such a productive environment and was close enough to markets, it might benefit more by producing tomatoes or cabbages than fuel. The relative benefit would change with market swings, of course, and the advantage would shift over time if other households began to seek the same market opportunities, if groups of households developed cooperative arrangements for water distribution or manure collection, if cabbage imports increased or fertilizer prices declined, or if the central government altered its subsidies or tightened access to the natural forest.

The sustainability of the fuelwood garden would depend upon what was happening in these many layers and scales of human activity, from the plot and household to the circle of surrounding villages, the market region, and broad relations between local and national powers and authorities in the forest. External factors shape what the fuelwood garden actually could become (Romm 1986; Thomas 1988; Govil 1990).

The fuelwood garden was intended initially to relieve household pressures on and to thereby sustain the natural forest. In what conditions might it do so? Adding population growth to the picture and unless firewood use became more efficient or was supplanted by other fuels, the productivity of gardens could not just be sustained: it would have to increase if this purpose were to be satisfied. Sustaining the natural forest required tremendous adaptation to external forces outside the natural forest boundaries. The more fixed were the desired characteristics of the forest, the more complex the social protection of the forest had to become.

The redwood forests of Northern California seem a world apart from fuelwood gardens in the Himalayas, but they are not so different when viewed in terms of the conditions for their sustainability. External forces have shaped the redwood forests more strongly than is acknowledged by any of the parties disputing their future. Rapid population and urban growth and strong competitive pressures for capital have been particularly forceful. Regional forest structure reflects long-term trends in capital costs, in alternative investment opportunities, and in the development of debt instruments; these trends determine the weight of the financial burden that the forest must bear. This burden reflects as well the shorter boom-bust cycles of the timber market and the longer term changes in market boundaries. The spatial distribution of the forest reflects the relative strengths of pressures to convert land from forest cover to pasture, to crops, and to residential settlement.

The redwood forest we see is shaped much less by silviculture than by the large external forces that determine its structure and extent. These forces drive choices that perhaps are rationalized as silvicultural technique. Conflicts between industrial and environmental interests about silvicultural choices miss the real issues. These conflicts continue to weaken mutual capacities to protect against and compensate for the larger forces that prevent all interests from achieving whatever vision of a sustainable forest any may hold. Sustainable forestry in the redwoods will involve processes that diversify the buffers and channels between the forest and the contextual forces that shape it.

TRADE AND COOPERATION: COOPERATIVE FORESTS AND COUNTY RULES

The qualities of a forest depend upon the degree of trade that exists among the diverse forest interests and upon capacities to enforce the outcomes of these trades. Few forest boundaries confine the influences of forest conditions on, or the actions of, people outside the boundaries who have some stake in these conditions. Forest conditions are as much an expression of the mutual exchanges and obligations among all forest interests as they are of the treatments chosen to serve one particular interest. Managing a forest for some specific condition requires modes of exchange and obligation that adapt to changes of interest and capacity among those who must participate in workable arrangements.

In the same region of Nepal, we sought to establish a plantation of fast-growing species on a brushy north slope of "national" forest. We fenced the area and began to clear strips through the brush. Hundreds of villagers attacked the work party, claiming that we were destroying "their" forest. We sat for several days of discussion on the ground at the forest edge.

In the course of the discussion, we learned that the villagers had sustained the brush for centuries as their source of coppiced fuelwood, livestock fodder, and green manure for their fields. We also learned something about how our professional forestry training had created an aesthetic preference, which could dominate rational considerations, for forests that were straight, tall, and unpeopled. The villagers learned that the spindly seedlings of acacia and eucalyptus might ultimately provide benefits their coppice would not and at least deserved a chance.

A mixed system was agreed upon, restricting the proportion of cover that was strip-cleared and replanted and forming rules of access to satisfy customary claims. The form, composition, and treatment of the forest came to reflect a de facto mix of rights and values that was sufficiently satisfying to both parties to sustain cooperative forest protection and the forest conditions it shaped. The sustainable forest expressed, biologically, economically, and aesthetically, the mix of interests and capacities that was needed to control influences on its condition. Whether it would last came to depend upon the continued willingness of the government to abide by the agreement regarding "its" forest. A decade later, the cooperative approach became the centerpiece of Nepali forest policy. Cooperation has emerged more slowly in countries where the formal owners of forests have had relatively greater power to prevent villagers' claims.

In Northern California's Mendocino County, clashes between industrial forest owners and grassroots environmental groups—in the courts, through statewide referenda, in local government, and in the forest itself—have reached paralyzing, everyone-loses intensity. The County Board of Supervisors commissioned a representative Forest Advisory Committee to recommend policies that would sustain the economic and environmental viability of the forest domain. After several years of great and contentious effort, the committee negotiated a compromise package of measures to slow forest fragmentation, to establish a system of long-term ownership management plans, to rebuild forest growing stock, and to limit activities that were thought to be detrimental to riparian, watershed, and wildlife habitat conditions (Mendocino County Forest Advisory Committee 1992).

Toward the end of the process, the industrial members of the committee withdrew their support from the compromise and sought relief through alternative channels at state and county levels of governance. The eventual outcome, and the sustainable forest qualities associated with it, depends upon the extent to which the county and the state choose to legitimize and enforce, modify, replace, or ignore the trades made in the negotiating framework of the committee. Meanwhile, villagers have resumed attacks on work parties of loggers in industrial old-growth stands.

COMPLEMENTATION AMONG LEVELS OF GOVERNANCE: PANCHAYAT FORESTRY AND COUNTY FOREST ZONES

We now understand that *sustainability* has different ecological content when assessed at site, ecosystem, landscape, regional, and global scales of analysis; forest conditions in a place express dynamic resolutions of these differences. Similarly, human interests and capacities differ among the levels of governance—ownership, local government, state, etc.—that affect forest qualities; these qualities depend upon the relations between the powers, interests, and resources the various governments represent and employ. Sustainable forestry is a process of adaptively meshing the comparative strengths and interests of different levels of governance rather than of identifying some ideal approach at one level (Romm and Washburn 1985, 1987; Romm 1987).

In 1978, the government of Nepal established panchayat forestry as the centerpiece of its forest policy. *Panchayat forestry* is a system of cooperative management between the Nepal Forest Department and the local governments, the panchayats, to create real interests in the conservation of forests among those who otherwise have reason to deplete them. The Nepal Forest Department is a relatively homogeneous and unified professional agency with specialized technical and financial resources. The panchayats are socially and ecologically diverse. As in most local government, they have a potential comparative advantage in the control of land use through the many sanctions people in a locality, to the degree that they are socially cohesive, are able to exert on one another. Panchayat forestry seeks arrangements of obligations and benefits that combine state expertise and local land control to sustain forests with mutually acceptable characteristics. ·

Despite uniform national policy, the nature of department-panchayat arrangements has come to vary in almost infinite degree with the diversity of conditions in which they have been negotiated (Ghimire 1989; Baker 1989). The role of the department relative to the panchayat increases with accessibility from towns and declines with the cohesiveness of panchayat members; the balance of power between technical expertise and land control in these arrangements varies along these spatial and social dimensions. Forest area and qualities reflect the balance of state and panchayat interests, resources, and powers in a place, shifting, for example, from exotic plantations toward indigenous species and styles to the extent that the panchayats predominate. The sustainable area, productivity, and composition of a panchayat forest depend upon a panchayat's access to government services and upon the cohesion of its people.

About when Nepal initiated its panchayat forestry policy, the state of California formed a Timberland Production Zone to reduce the conversion of private forests to nonforest uses. Land in the zone was to be taxed solely on its value for timber production, relieving tax pressures to convert forests to agricultural and residential uses. By 1978, unless owners had chosen otherwise, the state had zoned all ownerships larger than 160 acres and already taxed for timber production alone. Virtually all industrial forest land and most large nonindustrial parcels were placed in the zone by this means. Within broad state limits, county discretion then determined the zoning of and acceptable activities on the other half of the state's private commercial forest land, all of which was nonindustrial.

California's forest counties seem as diverse as Nepal's panchayats. No two counties pursued the same zoning strategy. Nevertheless, a general pattern emerged: metropolitan and hinterland counties zoned larger portions of eligible land than did counties in transition toward agricultural and residential settlement, where the conversion of forests was most rapid (Washburn 1983; Romm and Washburn 1985). The proportion of eligible nonindustrial land that counties placed in the zone displayed a U-shaped relationship with county population density ($R^2 = 0.69$) or, in reverse, with financial dependence on timber. Subsequent withdrawals from the zone have occurred entirely in U-bottom counties.

The Nepal and California examples make the same point in different ways. The area and content of a sustainable forest depend on the balance of interests, resources, and powers that different levels of government exert in that place. California applied uniformly light limitations on the counties and offered financially permissive opportunities for owners; the resulting zones, reflecting the diversity among counties and owners, differed significantly in forest scale, uses, and structures. Nepal provided active technical and financial services to panchayats but distributed these services unequally over space and among panchayats with different capacities to arrange and fulfill agreements; the selective consequences of state-panchayat arrangements were not unlike those of the differential state-county relations in California. In both settings, sustainable forestry is the capacity and interest to adjust combinations so as to better tap the resources that actual state-local relations could afford and to achieve forests that are valued as more productive by all who must sustain them in that state.

INVESTMENT: SOCIAL FORESTS AND FOREST ECONOMIES

Forests are sustainable when people have the capacity and interest to replace forest qualities they otherwise would deplete, to prevent avoidable losses, and to bank qualities they expect to become more scarce. Sustainability requires the social process of investment. It requires the generation of resources beyond the demands of immediate need, the means to capture and use these resources to attain future forest benefits, systems of exchange that secure the values of these benefits, and sufficient assurance that those who invest will be able to claim or distribute these values. Managing the forest to maintain some specific condition requires explicit attention to the dynamics of investment and the effects of different institutional conditions on them.

In the later 1970s, India initiated massive programs for participatory forest establishment and protection. Public reserved forests were being decimated; forest authorities had lost credibility; environmental and populist groups had become increasingly knowledgeable and effective in the forest policy arena. Financed by governmental and international grants and loans, tree planting was promoted on farms, along roads and canals, in towns, and on community and unreserved government lands. Private wood-processing facilities expanded; nursery, distribution, and marketing systems spread rapidly.

The states executed the programs in social forestry; forestry officials gained skills as community organizers and development workers. Different patterns and approaches emerged over time. In some areas, private-sector development of wood markets and urban-driven agricultural wage rises promoted starlike plantation forests growing outward along roads from towns. In other areas, states developed share arrangements with villages in reserved "public" timber or land, a new mode of agrarian reform, to secure local commitment to forest protection. Insurance schemes were formed to protect farmers against forest damages; credit programs let farmers use trees for collateral; cooperative agreements secured legal rights in and reinvestment funds for joint panchayat-state forest management. Forest departments undertook development activities—construction of roads, water systems, schools, and clinics; agricultural extension and credit—that were needed in different circumstances to increase forest investment.

Social forestry in India is entering another phase in which self-sustaining investment must gradually replace foreign and governmental subsidies. The sustainability of the forest depends on this social transformation, and lessons from earlier phases are being used to define its features.

First, social forests—the aggregations of all trees people plant, grow, or maintain—have come to be recognized as enduring expressions of people's investment choices; social forestry—the contemporary program of interventions in social forests—does not need to create these forests but to understand why people have done so and how to increase their forest investments effectively.

Second, social forestry professionals have recognized that their practical ad hoc responses to innumerable problems shared a common focus on improving conditions for forest investment; these responses afford an initial foundation for more unified investment strategies at farm, village, and regional levels.

Third, the effects of macroeconomic policies on people's investment choices and the forests these form are increasingly understood; rather than rely on direct interventions to the degree that they have in the past, the central and state governments will be increasingly able to use general policies—prices, wages, credit, and infrastructure—to shape the broad conditions for forest investment.

These features of the social transformation suggest a diversification of investment processes, operating on different scales, that together will mobilize more forest investment and achieve more productive levels of sustainable forests than could otherwise occur.

Although India's social forests have been expanding visibly from year to year, signs of the increasing investment in them, the forests of California's North Coast seem hostage to investment conditions that continue to reduce sustainable levels of forest productivity and forest-based economic enterprise. Growing stock and harvest levels on industrial lands are declining dramatically as both move toward future sustained yield at reduced levels. Forest-based employment is dropping even more rapidly. Industry is reducing the labor content of its products. Mill closures are common. Secondary manufacturing plants are arising near urban markets and labor pools outside the region rather than near its forest-dependent towns. Although growing stock is increasing on nonindustrial lands, harvests remain weak because nonindustrial forests are not linked to markets under conditions that satisfy most of their owners.

Structural rigidity rather than social transformation characterizes the forest economy of the North Coast. Strong sources of rigidity originate outside the region. Competitive national capital and commodity markets, for example, force local reductions in forest stock and simplifications of forest structure to levels that are inconsistent with any view of regional economic and environmental viability. Federal economic policies have encouraged investment in processing efficiencies relative to forest productivity; apparent consequences in the North Coast include (1) excessive processing capacity and (2) processing strategies that homogenize wood and diversify composition products rather than specialize means to exploit the forest's diverse qualities. These patterns run contrary to interests in retaining "natural" forests and rural livelihoods or even in mobilizing nonindustrial timber to keep mills alive. But the sources of rigidity are local as well as distant. A feudal culture in the region's timber industry discourages recognition of others' rights and ideas, while the contending culture of ecological fundamentalism provides little room for the economic surpluses that make investment and sustainability possible. Few partisans accept that frontier power and pristine forests both are gone forever and that no one's preferred forest is possible without social cohesion that sustains investment for all.

India's diversification of forest institutions is strengthening social cohesion and increasing forest investment without needing agreement on forest values. The social fabric of the forest is becoming flexible and resilient enough to spawn opportunities and secure people's interests without bending them toward a special version of truth. Indeed, the form of the sustainable forest in these conditions is substantially less predictable than in the institution-sparse North Coast but is relatively more productive in the longer term by any measure of value.

INNOVATION

Sustainable forestry requires not just an acceptance or understanding of change but also the urge to create new technical and institutional opportunities. Sustainability requires human creativity and enterprise above all. Innovative governments, corporations, communities, and nongovernmental organizations are advancing the sustainability of forests by the actions they take to solve problems and improve life in the future. The noninnovative lock us in toxic debates about whose "sustainability" is right and true. There are no right and true answers, only means to absorb and reflect many values with due justice for all in the actions we take. In the North Coast, new means are needed rather soon if subdivisions are not to replace forests of every persuasion.

SUSTAINABLE FORESTS AND SUSTAINABLE FORESTRY

The meaning of *sustainable forest* depends upon which forest attributes, activities, or effects are to be sustained, at what levels and over what area and time period, by which means, and for and by which people. In sum, it depends on human values. As there is no site on earth in which only one such definition expresses the human values that determine what people do there, the concept of the sustainable forest inevitably raises issues regarding relations among different forest attributes, activities, and influences; among different levels and scales of activity, territory, and time; among different preferred means; among different beneficiaries; and among different actors. Inherent in the concept is conflict among the value systems that underlie these differences.

In contrast, *sustainable forestry* is a regime of actions that sustains and enhances forest qualities amidst value conflicts that otherwise would weaken these possibilities. Sustainable forestry is a set of adaptive social processes (e.g., protection, trade, complementation, investment, innovation) that evolve to settle or surmount value conflicts in generally beneficial ways.

A large gulf separates sustainable forestry and the forestry profession. The profession developed to protect and manage forests that were thought to be defined by one system of values. The boundaries, attributes, and effects of the forest were treated as given and clear. The choice of interventions was presumed to be value-neutral. The forces of nature were assumed to predominate. The concept of the forest had no people. It should not surprise us that forestry, the source of the concept of sustainability, is perceived as a partisan in the conflict of values rather than as a process for its resolution.

People are everywhere, and we have learned that their effects on forests in the past were much more pervasive than the forestry profession had accepted. All forests are consequences of human as well as natural processes. The grand patterns of their types and places are as much dynamic expressions of society and culture as are settlement structures and characteristic art forms. Forests are an ecological record of resolutions among value systems of the past. Sustainable forestry as a professional enterprise faces no simpler a situation today, requiring new capacities to understand, adapt to, guide, and serve the dynamics of values that continue to shape these patterns and their effects. The challenge for innovation is formidable and very refreshing.

REFERENCES

Baker, J. M. 1989. A classification system for predicting the probable effects of community structure on the outcomes of social forestry policy. Master's thesis, University of California, Berkeley.

Ghimire, M. 1989. Distribution of authority for the conservation of forest resources: An analysis of the community forestry policy of Nepal. Master's thesis, University of California, Berkeley.

Govil, K. 1990. Why do farmers plant trees? A study of Uttar Pradesh, India. Ph.D. diss., University of California, Berkeley.

Mendocino County Forest Advisory Committee. 1992. Final report to the Mendocino County Board of Supervisors, Ukiah, CA, January. (Industry representatives on the committee issued an independent minority report in the same month.)

Romm, J. 1986. Forest policy and development policy. *Journal of World Forest Resource Management* 2:85-103.

————. 1987. Meshing state and local roles in California hardwood policy. In *Multiple-use management of California's hardwood resources*, ed. T. Plumb and N. Pillsbury, 416-426. Gen. Tech. Rep. PSW-100. Berkeley, CA: USDA Forest Service, Pacific Southwest Forest and Range Experiment Station.

Romm, J., and D. Seckler. 1979. Intensive fuelwood cultivation. In *Voluntary action*. New Delhi: Gandhi Peace Foundation.

Romm, J., and C. Washburn. 1985. *State forest policy and county land control.* Berkeley: Department of Forestry and Resource Management, University of California.

――――. 1987. Public subsidy and private forestry investment: Analyzing the selectivity and leverage of a common policy form. *Land Economics* 63:153-167.

Thomas, D. 1988. Village land use dynamics in Northeast Thailand. Ph.D. diss., University of California, Berkeley.

Washburn, C. 1983. County implementation of California's timberland production zone. Master's thesis, University of California, Berkeley.

14

Spotted Owls and Sustainable Communities*

Julie Fox Gorte
Office of Technology Assessment

A few years ago, to the apparent delight of the national news media, Yellowstone National Park burned up. For a few weeks, the evening news featured leaping flames, sooty fire fighters, and once a wonderful highway sign announcing the city limits of Cooke City, Montana, changed by some wit to read "Cooked City." Editorial writers and much of the American public denounced the so-called "let burn" policy of the Park Service, and one of Wyoming's senators announced that the fires were sterilizing the soil, making regrowth impossible for centuries. And then, on the next two anniversaries of the fires, the news stories of Yellowstone's remarkable rebirth appeared, right on schedule.

Yes, that was a parable. Rebirth after a seeming disaster is not only possible, it's normal. It works for socioeconomic as well as ecological communities and, as with natural "disasters" (the term we often apply to large, abrupt changes), the reborn community often looks different from the original. If we stretch the analogy a bit further, we also note that ecosystems that are considered healthy are those that undergo changes—sometimes slow and evolutionary, like succession; sometimes swift and dramatic, like wildfire or clear-cutting. Change is not necessarily destabilizing in natural systems, nor is it in human systems. In fact, attempting

* The views in this chapter are those of the author, not necessarily those of the Office of Technology Assessment.

to arrest the forces inducing change can just as surely destabilize a community as can an earthquake.

Trying to reduce or curtail some socioeconomic forces also generates powerful eddies and countercurrents. We have chosen, for example, to subject a great deal of our economic life in this country to market forces, which can, according to one of my more cynical college professors, mean that people can starve to death in a Pareto optimal way. The Japanese, for example, make more reliable, fuel-efficient, better-handling cars than we do and sell them at competitive prices. As a result of our choice of economic system, which places the interests of consumers above those of workers or producers in such cases, the American motor vehicle industry has shrunk and hundreds of thousands of autoworkers have lost their jobs.

Although Americans have grown accustomed enough to our economic system to think of the pressure of foreign competition as a force of nature, such a choice would have been unthinkable in Japan, or France, or Italy, or even (highly vaunted free trader though it is) Germany. There is nothing inevitable or irresistible about choosing to preserve threatened or endangered species like the northern spotted owl or the snail darter, either. If we choose to carry out the processes of the Endangered Species Act, we will have chosen, from a menu of mutually exclusive alternatives, one set of values over another.

One of the simplest ways to paint the debate is as a choice of jobs or owls, and there is a bit of truth in that. If protecting the owl is more important than maintaining timber output and jobs (at least, in the short run), it will cost about ten thousand to thirty thousand wood industry jobs. Even if the choice were that simple, however, weighing those jobs against the possible loss of a species would, because it is a case of incompatible values, cause endless disagreement and impassioned argument. As it is, however, the situation is far more complicated. Already the Pacific Northwest is faced with the prospect of inevitable contraction of employment in the wood products industry, in part due to the need to reduce timber harvests to a level that is sustainable in the long term and in part as a response to competitive pressures that have already caused diminishing employment (even with record production). The owl has become, pardon the pun, a stalking horse for powerful changes already afoot in the area. This is not to say that, since several communities in the Pacific Northwest already face disruption, we should be indifferent to additional increments. It does mean that the question of how to promote sustainable communities, in this instance, cannot be resolved by the God Squad or by repealing or amending the act. Sustainability, for communities in Oregon, Washington, and California and for the states themselves, will depend more on how they adapt and mitigate than on how deep they dig in their heels.

DEFINING SUSTAINABLE COMMUNITIES

A community is one of those things that defies precise description, but we usually know when we are looking at one. One way of defining community is administrative: a town, a city, a state, or a nation are all communities. The valence of intracommunity bonding is usually expected to be weaker for a larger administrative unit than for a small one, although this is not invariably so. Another way of defining community is by interest: there is an environmental community, for example, and an industrial community, religious communities, and others that span administrative boundaries. Communities may also be thought of in terms of something James Fallows (1989) called the radius of trust. "Except for psychopaths, everyone treats someone else decently. The question is how many people are classified as 'us,' deserving decent treatment, and how many are 'them,' who can be abused. When the radius of trust is small, the society is carved into tribes, castes, clans."

Most people belong to more than one community. In the case of the old-growth controversy in the Pacific Northwest, the different communities include the nation, states, counties, towns, private industry, environmentalists, public agencies, and perhaps more. The values of the nation—specifically, our democratic republic, wherein the expression of the public interest embodied in the Endangered Species Act (ESA) is presumed to be the will of the majority—are at odds with the wishes of another community: the private industry owners, managers, and workers whose livelihoods are vulnerable to disruption.

Sustainability is a simple concept, difficult to practice. Its meaning is "able to remain in existence," but what that means on the ground depends entirely on one's perspective. For some communities whose economic livelihood depends heavily on a sawmill or a plywood mill, sustaining current conditions is not possible. In part, this is because of the ESA, but it is also due to a host of other factors, including unsustainable historical harvesting from private lands, a competitive situation in wood products markets that probably precludes intensive timber management on all but the most productive lands, withdrawals of land from the timber base due to poor productivity, our desire for wilderness, and the like.

For the nation, on the other hand, sustainability depends on our ability to manage change. This is not feudal Japan or imperial China; we cannot lock ourselves away from the world (neither, incidentally, could they, in the end). Our economy survives only if it can adapt to the changing pressures inside, like the demographic shift toward a more elderly population, and from other nations, like international competition. The pressures are often self-inflicted, representing conscious choices. Our commitment to consumer sovereignty, free trade, and free markets is a key reason for the nearly total demise of the U.S. consumer electronics and commercial shipbuilding industries. The recent announcements of tens of thousands of layoffs by just two companies—General Motors and IBM—are also pertinent examples.

What makes the United States a viable community is not so much its ability to mitigate the pressures for change but the ability to roll with the punches. We are better off if, as a nation, we can find ways to help the workers, companies, and communities involved in wrenching changes to find new vitality than if we cut our principles to fit the current shape of the economy. We survive if we are more like Lowell, Massachusetts, than like West Virginia.

Lowell is enough outside Boston to consider itself a separate community, not a suburb. In the first half of the nineteenth century, Lowell's location on the fall line made it a magnet for the vibrant, growing textile industry; between 1826 and 1850 its population expanded from 2,500 to more than 33,000. By 1920, there were more than 112,000 souls in Lowell, and two-fifths of them relied on the manufacture of cloth for their livelihood. But 1920 was the end of an era for Lowell; the Great Depression and the migration of the labor-intensive textile industry to the low-wage South caused manufacturing employment to dwindle; as the rest of the nation recovered from the depression, Lowell did not. The revival did not come until the 1960s, with the birth and meteoric growth of the electronics industry. Specifically, in Lowell's case, growth was concentrated in office, accounting, and computing machines; electrical and electronic equipment; guided missiles and space vehicles; and precision and scientific instruments. An unconscious testament to the economic processes of renewal is the headquarters of Digital Equipment Corporation, a handsome brick building that once was a textile mill (Flynn 1988).

West Virginia's history is much like Lowell's except for the last chapter. Once a thriving center of the coal-mining industry, the shift from coal to other fossil fuels between 1948 and 1968, combined with technological advances in mining, eliminated nearly half of the coal-mining employment in the Appalachian coal region in just five years. The subsequent depression of the local economy has never really lifted, despite many efforts in later decades to improve economic conditions there (U.S. National Commission on Technology, Automation, and Economic Progress 1966).

Obviously, we want to be more like Lowell than Appalachia. Many things set them apart, not the least of which is Lowell's proximity to a center of population and economic activity compared with Appalachia's relative remoteness and much poorer educational attainments. When the displacements happened, there was very little in the way of public assistance to compensate or help communities and workers to bridge the gap between the old work life and a new one, but there are now. Economic development offices exist at both state and national levels to help communities develop or attract new businesses, and the Economic Dislocation and Worker Adjustment Assistance Act (EDWAA) helps workers laid off through no fault of their own. Neither of these programs, of course, is any guarantee of a smooth transition; even with the best of help and intentions some disrupted communities wither, and some workers never recover their old wages and benefits. The persistence of poverty and illiteracy in West Virginia, even after decades of special programs designed to promote economic development there, should make us cautious about inducing economic disruptions lightly. However, despite some superficial resemblance, the economy and culture of the Douglas fir region has little in common with West Virginia, and there is a good chance that the economy of the region can adjust to the impending change in industrial concentration.

PROGRAMS TO HELP DISPLACED WORKERS AND COMMUNITIES

In 1982, as a replacement for the Comprehensive Employment and Training Act (CETA) programs, the Job Training Partnership Act (JTPA) was passed and signed into law. Title III of the JTPA was aimed at helping displaced workers by providing a menu of adjustment services, including job search assistance, vocational skills training, on-the-job training, remedial education, and relocation assistance. Table 14.1 shows the amount of

TABLE 14.1 **JTPA Title III Allotments**

Program Year	Number of Program Terminees	Allotments ($ million)
1984	113,600	216.5
1985	149,692	222.2
1986	149,692	95.5
1987	129,984	195.6
1988	135,566	286.6
1989	139,642	278.6
1990	NA	463.6
1991	NA	527.0
1992	NA	577.0

money spent on the program in recent years. The JTPA was later turned into EDWAA, but with little basic change in its functions.

The program, in the last six years of operation, has served around 200,000 people each year, with the number rising somewhat erratically from 173,000 in program year (PY) 1984 to nearly 235,000 in PY 1989 (U.S. Department of Labor, Bureau of Labor Statistics 1989).

In PY 1989, 66 percent of the participants in the program had entered employment after their termination from the program, a slightly lower percentage than in earlier years and slightly better than the percentage of all displaced workers (including mostly workers who did not participate in title III programs) who were reemployed when they were surveyed. The average participant spent about twenty weeks in the program, and the average wage of those who found new jobs was $7.58 per hour, slightly less than the average hourly wage of all full-time employees in the United States in 1990 and 93 percent of the average hourly wage in manufacturing.

Performance was better than national averages in the three states most affected by proposed reductions in timber harvesting. In California, the entered employment rate of participants was 70 percent, down substantially from 78 percent in 1988 and 77 percent in PY 1987. The average wage of "graduates"—known in program lingo as "terminees"—was $8.25 per hour. Oregon looked like a clone; the entered employment rate was 70 percent and the average hourly wage was $8.41; in Washington, the entered employment rate was 71 percent and the average wage was $8.90 per hour. In all three states, wages were higher and entered employment rates were lower than in previous years. As far as we can tell, then, workers displaced in these three states have been relatively successful in finding new jobs, with or without assistance. Their new jobs are likely to be higher paying than the average displaced worker's new job, but probably pay less than the jobs they left (U.S. Department of Labor, Employment and Training Administration 1989, unpublished data).

All of this, of course, is dependent on the U.S. economy generally. During recessions or periods of sluggish growth, the displaced worker fares worse than otherwise, and we are now somewhere at the end or in the midst of a recession. Even during the worst of times, however, most displaced workers manage to avoid falling through the net; as of January 1984, at the end of one of the most severe postwar recessions, 60 percent of all displaced workers were reemployed, and 14 percent had left the labor force—which probably meant that they had a spouse's income or retirement income. Statistics are sobering, but it is worth keeping in mind that the majority of workers who are displaced find other ways to live and generate income that sustains them and others in their communities. Too often in this debate, we implicitly assume that lost jobs mean lost people and lost income forever; this is simply untrue.

Like workers, communities can recover from the shock of a major economic disruption. Not all of them do, and those that manage are apt to have to work hard to replace the lost economic activity or, like Lowell, are lucky. Diligence, for community planners or state planners, means employing many means to incubate, attract, or grow new economic enterprises. This can mean providing business incubators—physical facilities, along with some help in financing and providing financial and advisory services—to generate new small businesses; it can mean providing necessary training and education, tax breaks, or other incentives to attract businesses from outside the area to locate enterprises in the community. The federal government and most states have economic development programs. They vary greatly in quality, like dislocated worker programs; the worst of them are little more than raiders, who travel to Minnesota in winter to try to attract frozen entrepreneurs to warmer climates, or carnival barkers, who try to outbid other communities for the newest Japanese firm to invest in the United States. Good programs, however, can save communities.

There is very little in the United States to help firms that fail. Every now and then, with great protestations, the federal government "bails out" a faltering firm—Lockheed in the 1970s and Chrysler in the early 1980s are examples. Usually this takes the form of loans or loan guarantees, along with various other regulatory concessions as needed. Can the forest products industry expect anything like that? It is most unlikely, for several reasons. For one thing, Lockheed and Chrysler only cost money. A bailout of the forest products industry in the Northwest would cost a scarcer coin: principles. Over the years, we have committed to sustained yield and the protection of endangered species, and we would have to compromise both to avoid the job losses impending on the West Coast. Even then, it is quite likely that employment would decline, for the same reasons that it has already: substitution of other materials for wood in many end uses and technological change, with its attendant improvement in labor productivity. As a result, we might have to give up too many of the things that define us ethically to save too few jobs. At the moment, too, environmental principles are worth more, on a pound-for-pound basis, than many others; President Reagan's attempt to dismantle much of the environmental regulatory infrastructure was a debacle, based on a misreading of American sentiment. The Bush administration slowly learned the same lesson.

More to the point, however, is the fact that the forest products industry would be ill-advised to rely on some form of bailout because it has no Lockheeds and no Chryslers. Lockheed was important because, as a prime contractor to the Department of Defense, it had a unique ability to develop and make weapons, particularly advanced aircraft. Considering the price tag for developing aircraft—running in the billions—Lockheed represented a singular resource, one that the Air Force could scarcely replace. In the case of Chrysler, roughly two hundred thousand jobs were at risk if the company failed. Finally, Chrysler's bailout ended up costing exactly nothing: it was a loan guarantee, and the company never defaulted. True, it did represent a compromise of our preference for laissez-faire approaches to market regulation, but this was a compromise that, unlike the case of the northern spotted owl, involved no irreversible losses.

CONCLUSION

The economic life of the Pacific Northwest is going to change in the next decade. Almost certainly, Oregon and Washington will be less dependent on wood products manufacturing than they are now, and at least ten thousand jobs will disappear. This will threaten community stability, but it need not do lasting damage to the region or to more than a handful of smaller towns. The damage will be greater if those most involved in the dislocation—the industry, its workers, the communities, state and local civil servants, the federal government, and the Forest Service—are unable to mobilize resources to aid the victims and rebuild. There are some resources available in the form of federal and state programs to aid workers and communities.

One other ingredient is sorely needed: the will to adapt and adjust. Although not quantifiable, the will to pick up the pieces and go on is the closest thing to a silver bullet there is. We can see this will in action if we consider, for example, the recovery of the devastated economies of Japan and Germany after World War II. In both cases, with a lot of help from their new allies and a deep well of grit and determination, these communities were able to recover, not just well, but spectacularly well. Thus far, however, few of those most involved in the timber controversy in the Pacific Northwest have been able or disposed to consider life after cut-backs, focusing instead on the projected magnitude of the loss and, in some cases, on ways to stave it off, most of them probably futile. If we wish to foster a sustainable economy in the region, it is time to move on, time to build new economic enterprises.

REFERENCES

Fallows, J. 1989. *More like us.* Boston: Houghton Mifflin.
Flynn, P. M. 1988. *Facilitating technological change: The human resource challenge.* Cambridge, MA: Ballinger Publishing.
U.S. Department of Labor, Bureau of Labor Statistics. 1989. JTPA Title III analysis. Unpublished.
U.S. National Commission on Technology, Automation, and Economic Progress. 1966. *Technology and the American economy,* vol. 1. Washington, DC: US Government Printing Office.

15

Leadership and the Individual: The Keys to Sustainable Forestry

Robert T. Perschel

Institute for Environmental Leaders

The theme of this book is defining sustainable forestry, but we are also seeking something much larger and more expansive than that. We are looking for the leadership needed to achieve sustainable forestry. This chapter discusses leadership and the relationships among leadership, the individual, and change.

To achieve our goal of sustainable forestry, we must forge a new relationship with the land, and this chapter addresses our capacity to lead this change. Leadership results when the powerful forces within our hearts are rallied to seek and realize the highest values of which we are capable. When we are aware of and direct these forces to attain our goals, meaningful change occurs, and the impossible becomes a reality.

The 1992 conference on defining sustainable forestry demonstrated the beginning of leadership within the profession. We came together to seek an ideal that we could not yet define. This is a brilliant example of an essential quality of leadership—the desire to produce things greater than we have yet imagined. With full awareness of the historical achievements of forestry, we wish to produce something more expansive. Economic productivity, huge volumes of wood, and incredible numbers of regenerated acres of trees are not enough. Something is missing, and we have begun to notice it. As impressive as those achievements are, we know that we are capable of more. Something within us wants to go beyond what has been expected of us, beyond what has been accepted from us, and beyond what we have produced to this point. We wish to create a new relationship between humankind and the forest, a relationship that we do not yet have words for but that is already a reality within our hearts. For many of us it has been present since the first moment we stepped into a forest. Now it is time to lead our country to this reality.

To accomplish this we must understand leadership as have no professionals before us. We must become aware of and utilize the power of the individual to create change. Working with environmental professionals on ethics, change, and leadership, I have seen the demonstration of a universal principle. Any change for the betterment of humanity must begin with the individual.

Consider this principle in light of what transpired at the conference. We recognized and defined some of the vital changes that must take place in organizations, institutions, and society to achieve sustainable forestry, but it was more difficult to describe how these changes are to be effected. We endowed these organizations and institutions with energy, will, and life force. Although this may seem to be true, it is only because individuals set these forces in motion. The fatal mistake is to assume that institutions and organizations operate independently of the individual. They were created by the energy and will of individuals, and they must be recreated from this same source. If an institution stagnates and is irrelevant, it is because of the individual. If it is vibrant and adaptable, it is also because of the individual. Foresters, like any other professionals, often fail to respond to the drive for leadership, and we must examine these failures. We must ask ourselves if we are capable of leading this country to sustainable forestry. We must be aware that we hold within us the capacity to stifle growth and change as well as the capacity to bring others to achievements they did not think possible.

Leadership is a universal principle and an instinctual drive of humankind. Confusing leadership with an individual or a particular style, such as charismatic or transformational, is to confine ourselves helplessly to the limits of these individuals and the prescribed styles. Each one of us is capable of uniquely expressing and demonstrating creativity, intelligence, and the drive to achieve and draw others to a compelling and sustaining vision for humankind and the earth.

America is an example of the leadership principle in action. Our Constitution offers the autonomy and freedom needed for the individual to achieve in accordance with the highest standards of the human spirit.

These gifts come with the imperative that we attain these standards in all aspects of our lives. In other words, we must be leaders, now and forever.

In this country we have great freedom with our forests. The Constitution gives private citizens the right to care for or abuse the land. If individuals respond by seeking their most meaningful values, then we, as Americans, have the greatest potential on earth to achieve sustainable forestry. Our wilderness system, our national forest system, and the Endangered Species Act attest to this potential. Conversely, if we choose to pursue anything less, we will fail. Past and present abuses of our land attest to this. Initially, our failure may occur less as a loss of quantifiable environmental quality within our own country than from the worldwide consequences of the absence of our leadership.

There is an obligation to lead whether we are considering Pacific Northwest old growth or the regenerated hardwood forests of the East. For the first thirteen years of my career, I worked in the forests of New England. On a daily basis I met private landowners who often were not committed to caring for their land. Financial analysis and operating economic models encouraged the abuse of highgrading. Local laws did little to support true forest stewardship. Incentive programs were paltry. Many landowners were urbanites, disconnected from the cycles of the land. After thousands of experiences and countless forestry conferences seeking solutions, I began to ask if there was something else that might be effective in changing the ways in which landowners cared for their land. I studied leadership and change for three years, seeking counsel from experts in the field—people who worked to cause change in corporate America, government, and educational institutions.

I discovered that the change I wished to see take place on the land had to begin with me. I had to become the change. When I stood firmly upon what was most meaningful to me and did this with utmost integrity, people began to change in the way they treated their land. Not all landowners became land stewards, but most began to reach for something greater and committed themselves to better forest management. Other foresters, when exposed to this principle of leadership in action, responded in kind, as if they were waiting for this imperative.

Leadership, when manifest through an individual, impels the same principle to emerge in others. This implies that leadership can cause sweeping changes in organizations and institutions. It also explains why we languish in so many areas critical to our society such as education, technology, and business competitiveness. Leadership must ignite within some one individual or, better yet, within several or many for real change to occur. If all individuals merely stand around looking for a leader, nothing happens. If you feel that leadership is necessary for change, then become the change, become a leader.

We need to nurture and develop leadership within our profession. This work must be direct, personal, sustained, and focused on activating the individual's drive to lead. It does not fit efficiently into the one- or two-day workshop or seminar format. Two major forestry leadership development efforts over the past year illustrate this.

Last summer my colleagues and I were invited to address the House of Society Delegates of the Society of American Foresters concerning the inclusion of a land ethic in their code of ethics. During a two-hour workshop all participants described their first recollections of the natural world. One after the other they stood at the podium and delivered the most poetic, emotional, and sensitive descriptions imaginable. Their memories covered the entire country—Alaska, the Olympic Peninsula, Texas, and Florida. They were truly magnificent describing their original connection to the land. The energy and love through which they had been drawn into the forestry profession were tangibly present, yet the experience was fleeting. When they were asked to articulate a land ethic, this energy faded as if it were not real or appropriate to consider in the worldly pursuit of a professional society's ethic. Many participants noted this failure with disappointment.

The inability to sustain this wonderful connection to the natural world is a tragedy within the forestry profession and our society. Each of us has an experience of this failure. A colleague, Dr. Marie Loscavio, put it best.

> Like others you chase those moments in your vacations and vacation dreams, in your photos and paintings, in poems and songs, in beautiful dreams and memories and endless stories told trying to capture and express, trying to hold that which for one brief and compelling instant succeeded in capturing you for all time.
>
> Why is it that every time you have an experience that is filled with the joy of being you also have the complete dumbness of having to name it other than reality?

We can no longer fail to embrace this reality. The future of our forests and our world hangs in the balance. It is time to call forth the ever-present energy and creativity that come from our love of the forest.

Although most leadership programs emphasize intelligence and knowledge, a leadership development program for foresters must go further and include the intuition and emotion that we derive from our intimate connections to the land.

In another leadership effort I served as an advisor to a group of eight foresters chosen to develop a new mission statement for the New England Society of American Foresters. The result was a mission far more expansive than any had thought possible. This success was a direct result of the method used in the sessions. The group decided to produce the most expansive mission possible and not to limit themselves to a statement designed to pass a vote. The meetings were sustained over more than six months. Each individual explored his or her relationship to the land and was continually challenged to reach for something greater. Although people did not agree on every aspect of the new mission, there was unanimous agreement that the joint product was the most expansive vision that they could produce and live up to. Leadership was sustained through the entire mission project and today, as these foresters practice their profession, they continue to draw on that experience in pursuit of their mission. They are capable of leading others to achieve sustainable forestry.

In conclusion, if we are to achieve sustainable forestry the role of leadership must be understood and then developed through a concerted effort. We must recognize that the individual is the key to change, that leadership is instinctual and demands an individual response, and that leadership impels similar responses from others. Our best hope at this time is to turn our attention and our resources toward the individual to tap the elemental energy that can assure our success. Discovering and listing another 1.4 million species will help only if we can provide the leadership to care for the 1.4 million we already know.

We are offered a great opportunity. When some day those who follow us look back at the period of years that spanned our careers as stewards of the earth, will they see a time in which the most rapid extinction of life in history was reversed and diversity and complexity again increased? Will they see a time when vibrant leadership halted the decline of the quality of life on this planet and replaced it with a new reality forged from the desires of our hearts?

I am reminded of a poem by Archibald MacLeish called "An Epistle to Be Left in the Earth." In it he explains all that we have learned of the earth and asks those who would, millennia from now, find this wandering earth and open his writings to "make in your mouths the words that were our names."

What legacy will we leave during our brief, yet monumental, time as stewards of this earth? For those who believe that we are capable of producing something greater, there is only one response—leadership.

CONCLUSION

Prospects for a Sustainable Future

Gregory H. Aplet

Nels Johnson

Jeffrey T. Olson

V. Alaric Sample

The watchword for the 1990s has been "change." This has been true not just of campaign oratory, but of discussions of every topic from international affairs to urban policy. Forestry, too, has undergone significant reform. In June 1992, USDA Forest Service Chief Dale Robertson announced a historic shift in Forest Service emphasis from commodity production to ecosystem management. The Bureau of Land Management announced a similar program of "Total Forest Management." These changes may eventually be recognized as the most significant events in the history of forest management since the creation of the Forest Service.

But change in forestry is being felt at a more fundamental level than mere political rhetoric. A century of ecological research has resulted in recent widespread recognition of the pervasiveness of change in ecosystems. We now recognize that ecosystems at all scales do not and cannot remain static. This has forced upon us a new paradigm of forest management. We can no longer attempt to manage for the optimal *state* of the forest. We must see the forest for the dynamic, constantly changing entity that it is. Rather than battle against change, we now know we must work with it. Achieving sustainable forestry will require the political act of changing forestry to respond to changing ecosystems.

Throughout this book, no single theme has been better developed than the idea that sustainable forestry must be ecologically sound, economically viable, and socially desirable. (Some would add a fourth, politically acceptable.) These constraints have been represented as equally important; take any one away and forestry will not be sustainable. Although this is undoubtedly true, it is important to keep in mind that forestry is a human construct; it is about making forests perform for people. Thus, the decision to engage in forest management is a social/political/economic decision from the outset. Ecological soundness reflects human values; without management, forests may self-perpetuate, change character, or progress from a forested to a nonforested condition. This behavior is value-free until we impose our desires on it. We engage in forestry to guide ecological processes to create or maintain forests of a desirable quality. Despite the contentious nature of the present debate, one quality—sustainability—has emerged as universally desirable.

Almost fifty years ago, Aldo Leopold (1949), in his essay "The Land Ethic," discussed an ideological rift in resource management that he called "the A-B cleavage." Sadly, his words are as relevant today as they were when he wrote them.

> In my own field, forestry, group A is quite content to grow trees like cabbages, with cellulose as the basic forest commodity. . . . Group B, on the other hand, sees forestry as fundamentally different from agronomy because it employs natural species, and manages a natural environment rather than creating an artificial one. Group B prefers natural reproduction on principle. It worries on biotic as well as economic grounds about the loss of species like chestnut, and the threatened loss of the white pines. It worries about a whole series of secondary forest functions: wildlife, recreation, watersheds, wilderness areas. To my mind, group B feels the stirrings of an ecological conscience.

Traditionally, resource management, in the United States at least, has been dominated by group A. The primary objective of forest management has been fiber production, and even the management of wildlife, fisheries, and range has focused on similarly narrow outputs. The firm influence of group A is evident in the pervasiveness of such concepts as "full stocking" and "culmination of mean annual increment" throughout forest management. These are agronomic terms more associated with volume production than with the management of a natural environment. The forest produced by group A consists of fully stocked stands of young, healthy, rapidly growing trees—a tree farm.

The heightened debate over the future of forestry reflects public dissatisfaction with the tree farm as forest. A forest is not a perennial row crop. It is, as Leopold said, "something broader." It is a complex, natural, ecological community. To an increasing number of people, a forest that is not "natural" is not sustainable; a forest that has lost natural composition, structure, or function is not sustainable. Group B understands this, and the outcry for sustainable forestry can be interpreted as a public demand that group B take charge of forest management.

The hallmark of management under group B is the condition best described as "naturalness." Naturalness is a desired condition, not simply because of its aesthetic appeal, which should not be trivialized, but because it represents the highest likelihood of achieving Leopold's most famous objective: "to keep every cog and wheel." Jerry Franklin (chapter 4) described the same goal when he wrote of sustainability as maintaining the *potential* of the ecosystem. Because we have only recently begun to understand the complexity of interdependencies in forest ecosystems, we cannot hope to "design" a forest that satisfies anything but the most basic economic objectives. If we want anything but cabbages, we must rely on nature as the model of the sustainable forest.

The best model we have of this functional, natural forest is the North American landscape before alteration by Europeans. This landscape, while certainly undergoing constant change, was the product of forces that remained relatively consistent over several thousand years of human habitation. It was these "presettlement" landscape dynamics that maintained the composition, structure, and function that we recognize as naturalness. Although conditions, particularly human population density, have changed dramatically, presettlement landscape dynamics can provide useful models of sustainable ecological systems and have already contributed to the development of concepts of ecosystem management (see chapter 5). Further development will depend on similar applications of natural processes to forest management in all regions.

Managing for naturalness, though, presents a paradox. How can we have resource extraction and natural processes when the removal of even one log compromises those processes? The answer lies in the word *compromise*. The future of sustainable forestry lies in finding the level of extraction that also perpetuates the natural dynamics of the forest. That is, we must identify the forest that is *natural enough*. This notion is captured in Cairns's definition of biological integrity as "the maintenance of the community structure and function characteristic of a particular locale or deemed satisfactory to society" (Cairns 1977). Thus, sustainability carries with it components of both functioning native ecological systems and human desires.

Perhaps this is unachievable. Perhaps, through further research, we will learn that extraction and naturalness are incompatible. This is undoubtedly true for very fragile sites or complex, ancient communities, but it may also be the case for forest lands in general. If so, a mix of extractive resources and naturalness may be achieved only on separate estates. Society would be forced to choose how much forest it wants to keep forest, and how much it wants to convert to farm.

Unfortunately, there remain precious few places on earth where natural forest can be carved off and managed for pristine conditions. Where these forests exist, some significant proportion of them should be maintained for values not obtainable from the managed landscape. These lands provide unique recreational, spiritual, and other values, but they also serve as essential reference points against which to judge success on the managed portion. Functioning natural forest should be maintained as a useful "blueprint" for management.

For the vast majority of this planet, however, we no longer have the luxury of dominant-use allocations. We must find ways of maintaining and utilizing the full array of values that forests provide. We must discover and maintain the forest that is "natural enough." This challenging new approach will require recognizing the variability in natural forest ecosystems and working within that variability to achieve economic/social/political objectives. In this sense, ecology is not an equal partner with economics and social acceptance. Rather, it sets the constraints or defines the decision space within which human objectives can be achieved.

One application of this concept has been developed in the USDA Forest Service's Northern Region and has recently been applied to a difficult management challenge in the Blue Mountains of northeastern Oregon. The approach was originally called *Sustaining Ecological Systems* (SES) but is now the Northern Region's model for Ecosystem Management. It relies on identifying the range of vegetation conditions present in the preindustrial forest and using those conditions to guide management. Once the range of "natural" conditions is defined, current conditions can be compared, and treatment can be recommended to bring future conditions in line with natural composition, structure, and function. The approach is being applied in the Blue Mountains to assess how best to direct management efforts to alleviate undesirable forest conditions resulting from fire suppression and unsustainable logging practices (Blue Mountains Restoration Panel 1992).

The approach being used in the Blue Mountains is an attempt to identify and manage for a forest that is natural enough. It relies on identifying the level of extraction that can be sustained without destroying the integrity of the ecological system. Because there is no true waste going unused in a natural system, such an approach depends on the existence of some level of redundancy or "slop" that can be utilized. This is likely to be considerably less than net volume growth. Identifying the sustainable rate of removal represents one of the greatest challenges to sustainable forest management.

Another great challenge to forestry is to reconcile the management of mixed ownerships. Property lines rarely follow ecologically meaningful boundaries. Management of ecological systems will require the cooperation of multiple landowners with various objectives to achieve common goals at the landscape level. New, cooperative institutions must be developed that facilitate cooperation among various owners to achieve disparate goals within common ecological constraints.

Finally, new technologies and new policies must be developed that promote the efficient use of the products that are removed from the forest. Wood is a vital, environmentally sound product whose use should be promoted in national efforts to reduce energy consumption and close the global carbon cycle. In addition, bold efforts must be taken to reduce unnecessary consumption and avoid the exportation of U.S. environmental problems. As a society, we must strive to balance domestic wood product consumption with a domestic supply capacity. A strong demand management policy is an essential part of the sustainable forestry equation.

Defining sustainable forestry requires resolving what society wants from its forests at various scales. From rural towns to the nation as a whole, people must be involved in the development and implementation of sustainable forestry concepts. As in the past, the best "scientific forestry" will fail to produce workable solutions unless society stays actively engaged in the debate. If the process is opened for participation of individuals at all levels, a healthy proliferation of approaches will result. Just as various ecosystem elements and processes emerge at different scales, so may various models of sustainability. Clearly, there is a need to restore naturalness to vast areas of the landscape. However, this should not preclude intensive fiber production where this does not have significant negative effects on regional sustainability. In the end, the multiplicity of relevant scales may serve rather than hinder progress toward sustainable forestry.

In conclusion, these are exciting times for forest management. Change is happening fast. Upheavals in management from the Pacific Northwest to the Coastal Plain signify the beginning of a new era in forestry. It is our hope that the ideas presented in this book will help catalyze continued explorations of the meaning of sustainable forestry.

REFERENCES

Blue Mountains Restoration Panel. 1992. *Restoring ecosystems in the Blue Mountains: A report to the regional forester and the forest supervisors of the Blue Mountain Forests.* Portland, OR: USDA Forest Service Pacific Northwest Region.

Cairns, J. 1977. Quantification of biological integrity. In *The integrity of water*, eds. R. K. Ballantine and L. J. Guarraia, 171-187. Washington, DC: U.S. Environmental Protection Agency, Office of Water and Hazardous Materials.

Leopold, A. 1949. *A Sand County almanac.* New York: Oxford University Press.

About the Contributors

🌿

Gregory H. Aplet is forest ecologist in The Wilderness Society's Bolle Center for Forest Ecosystem Management.

Jerry F. Franklin is the Bloedel Professor of Ecosystem Analysis in the College of Forest Resources, University of Washington.

John C. Gordon was dean of the School of Forestry and Environmental Studies at Yale University and is now Pinchot Professor of Forestry.

Julie Fox Gorte is a senior associate in the United States Congress's Office of Technology Assessment.

George Honadle is a consultant and researcher at Hidden Creek Farm, North Branch, Minnesota, and Adjunct Professor in the Department of Forest Resources, University of Minnesota.

Nels Johnson is an associate at the World Resources Institute, where he works on forest and biodiversity policy research.

Steven M. Jones wrote his chapter while on the faculty of the Department of Forest Resources at Clemson University. He is now a consultant at Environmental Services, Inc., in Atlanta, Georgia.

F. Thomas Lloyd is a researcher at the USDA Forest Service's Southeast Forest Experiment Station, Clemson, South Carolina.

Douglas W. MacCleery is assistant director of Timber Management, USDA Forest Service, Washington, D.C.

David J. Mladenoff is a forest ecologist in the Natural Resources Research Institute, University of Minnesota, Duluth.

Reed F. Noss is an independent consultant in conservation biology and a part-time research scientist with the University of Idaho, College of Forestry, where he works on the "gap analysis" project of the U.S. Fish and Wildlife Service.

Jeffrey T. Olson is director of The Wilderness Society's Bolle Center for Forest Ecosystem Management.

George R. Parker is a professor of forest ecology in the Department of Forestry and Natural Resources at Purdue University.

John Pastor is a forest ecologist in the Natural Resources Research Institute, University of Minnesota, Duluth.

Robert T. Perschel wrote his chapter while directing the Institute for Environmental Leaders. He is now Northeast Regional Director for The Wilderness Society in Boston, Massachusetts.

Robert D. Pfister is research professor of forest ecology in the School of Forestry at the University of Montana.

Alice M. Rivlin wrote her chapter while senior economist at the Brookings Institution.

Jeff Romm is professor of forest policy in the Department of Forestry and Resource Management at the University of California, Berkeley.

Hal Salwasser wrote his chapter while director of New Perspectives in the USDA Forest Service. He is now the Boone and Crockett Professor of Wildlife Conservation in the School of Forestry at the University of Montana.

V. Alaric Sample is director of the Forest Policy Center at American Forests in Washington, D.C.

Thomas A. Snellgrove is branch chief of the Forest Products and Harvesting research staff of the USDA Forest Service in Washington, D.C.

William D. Ticknor heads W. D. Ticknor, Forestry Consultants, Inc., in Orient, Ohio.

Michael A. Toman is a senior fellow at Resources for the Future in Washington, D.C.

Edward O. Wilson is Frank B. Baird, Jr., Professor of Science and Chief Curator of Entomology, Museum of Comparative Zoology at Harvard University.

Index

Island Press Board of Directors